Money Trees

Money Trees

*The Douglas Fir and
American Forestry, 1900–1944*

Emily K. Brock

Oregon State University Press
Corvallis, OR

Library of Congress Cataloging-in-Publication Data

Brock, Emily Katherine, 1973-
  Money trees : the Douglas fir and American forestry, 1900–1944 / Emily K. Brock.
      pages cm
  Includes bibliographical references.
  ISBN 978-0-87071-809-0 (original trade pbk. : alk. paper) — ISBN 978-0-87071-810-6 (e-book)
  1. Douglas fir. 2. Douglas fir—Economic aspects—United States. 3. Lumber trade—United States. 4. Forests and forestry—West (U.S.) 5. Douglas fir—United State—History—20th century. I. Title. II. Title: Douglas fir and American forestry, 1900-1944.
  SD397.D7B74 2015
  338.4'7674—dc23
                              2014046157

Oregon State University Press
121 The Valley Library
Corvallis OR 97331-4501
541-737-3166 • fax 541-737-3170
www.osupress.oregonstate.edu

To Brian,
brother and friend

Vancouver
Island

Strait of Juan de Fuca

Cape Flattery

C A N A D A

W A S H I N G T O N

Seattle

Tacoma

Spokane

Columbia River

□ *Clemons Tree Farm*

PACIFIC OCEAN

□ *Wind River Expt. Sta.*

Columbia River

Portland

Willamette River

Eugene

Bend

O R E G O N

Coos Bay

Cape Blanco

CALIFORNIA

NEVADA

─·─·─·─ Country boundary	Historical range of coastal Douglas Fir
───── State border	0 km 200

Contents

Acknowledgments

Thanks to Andrew C. Isenberg, patient adviser and stalwart friend. Thanks to Richard White, Nancy Langston, and David M. Kennedy for advice and inspiration. Thanks to Mary E. Braun and to the wise anonymous reviewers at Oregon State University Press.

I received several fellowships and grants that aided me at various stages of the research and writing process, including a postdoctoral fellowship from the Bill Lane Center for the American West at Stanford University, the Sterling Senior Fellowship in Pacific Northwest History from the Oregon Historical Society, the Alfred D. Bell Research Fellowship from the Forest History Society, and the Grey Towers Scholar-in-Residency from the US Forest Service and the Grey Towers Heritage Association. Finalization of this manuscript took place in the Philippines during a Fulbright Foundation Senior Research Fellowship, and in Germany at the Rachel Carson Center for Environment and Society in Munich and the Max Planck Institute for the History of Science in Berlin. My thanks for the generosity of those programs and the kindness of those who administer them. My thanks also to the University of South Carolina and Georgia State University for their generous support.

The ranks of scholars whose conversations, perspectives, corrections, and encouragements I have valued are too voluminous to list entirely. Special thanks, however, to James Morton Turner, Joanna Dyl, Lissa Wadewitz, Jeremy Vetter, Darren Speece, Lee Worden, Sara Schwebel, Katja Vehlow, Michael Rawson, Nick Wilding, Joshua Howe, Zena Hitz, Ari Kelman, Dagmar Schäfer, William G. Robbins, and Philip Pauly.

My thanks to those faculty who nurtured my scholarship and creativity in the Princeton University History Department, especially Anthony Grafton, Angela Creager, William Jordan, and Gerald Geison, as well as ElizaBeth A. Fox, Henry Horn, and my fellow graduate students who have become treasured friends. Thanks also to those who taught me how to be an ecologist, then helped me find my way from grad school in ecology to grad school in history: Russell Lande, John H. Willis, T. F. H. Allen, Warren Porter, Emília Martins, Nancy Tuana, and Russell McCormmach. And thanks to my earliest intellectual mentors, Erik Sageng and Henry Higuera, who showed me how to read boldly.

Without archives, historians would be adrift. Thanks to all the kind, welcoming, and resourceful archivists and librarians I have encountered during my research, especially Cheryl Oakes at the Forest History Society, Lori Williamson at the Minnesota Historical Society, and the sunny staff of the National Archives at Seattle.

Thanks to S. Tim Yoon, Sarah Madden, and Chris Koerner. Thanks to all my dear friends for making this life fun. And my deepest thanks to all the Brocks for their love, support, and everything else.

Tract of "virgin forest" near the Wind River Forest Experiment Station, in the Columbia National Forest between the Columbia River and Mount St. Helens (1935). The trees present when this photo was taken were mainly Douglas fir with some hemlock. In the absence of any fire, logging, or other disturbance, the hemlock would gradually increase in proportion as the forest aged. Photograph 312839; Records of the Forest Service, RG 95; National Archives at College Park, MD.

Introduction

Fed by plentiful winter rains and nourished by a mild summer sun, lush temperate rain forests flourish on the Pacific edge of North America. From the cool fog of far northern California, stately Douglas fir ranges north toward Alaska and spans from the western slopes of the Cascades to the ocean. Coast Douglas fir (*Pseudotsuga menziesii* var. *menziesii*) forests are some of the richest in the United States, both ecologically and economically. American lumber companies began to log these forests in an organized and concerted fashion around the turn of the twentieth century. The wood proved so lucrative that the tree took on new value and meaning for the regional lumber industry, allowing it to grow at a furious rate. The early twentieth century's massive harvest of the Douglas fir did not happen mindlessly; behind it lay specific decisions made by professional foresters intent on controlling the flow of lumber and the forest itself. These foresters were employed by the government, the logging companies, and the forestry schools of American universities. Despite their academic training, they often worked with very little knowledge or expertise about an undeveloped and little-understood type of forest. The profession of forestry was itself forced to evolve to accommodate the fast-growing Douglas fir industry.

The histories of American forests and professional forestry have been locked together for more than a century. The first true foresters in the United States were Europeans who arrived in the late nineteenth century, but the profession took root domestically soon after. As the nation's lumber activity centered in one or another region, the focus of forest management moved with it. Loggers, empowered by new clear-cutting technology, worked quickly in the Douglas fir, leaving behind denuded hillsides. As the lumber industry of the Northwest became increasingly lucrative, managing logging in that region became a central focus for American foresters. Every ecosystem is unique and presents foresters with different challenges. In the case of the Douglas fir, its unusual attributes would include its dominance as the main species in the forest canopy, unusual physiology, and immense size and height at maturity. The almost impenetrably rugged terrain in which the forest flourished posed challenges too. These Douglas fir forests would prove both interesting to scientific researchers and challenging to loggers. The nascent community of American professional foresters struggled to learn all it could about the Douglas fir while also keeping control of its rapidly escalating harvest. As foresters worked to manage the forest, the federal structure of public forest management, and indeed the very profession of forestry, evolved. An entity that began the century as the small, loosely organized, research-centered Bureau of Forestry became the massive, centrally controlled, economically centered US Forest Service of midcentury. Logging changed too: the cut-out-and-get-out logging concerns of 1900 turned into the Tree Farms of the 1940s. The changes on the ground in the Douglas fir reflect the shifts in how foresters did their work and planned for the future.

When one walks through a Douglas fir forest today, the traces of the past are evident all around: fire scars, abandoned rail spurs, roadbeds built with New Deal dollars, grid-like stands of second growth. In a place where each tree lives so long, decisions made a lifetime ago may still haunt the forests of the present. To understand why the contemporary forest looks as it does, we need to delve into its past.[1] But the freewheeling heyday of Douglas fir logging did not involve just cutting down trees. American professional forestry was shaped by both the ecology of the Douglas fir forest and the realities of its harvest. In every place and era, forestry is grounded in local environmental realities, even as it simultaneously focuses on national—even global—trade and policy. If we want to know why the forests of today look as they do, we must examine the careers of those who set the agendas for American forest management.

Aldo Leopold and What We Think We Know about Forestry

One of the most studied, and most revered, foresters from the first half of the twentieth century is a man who famously rejected forestry altogether. The former US Forest Service employee Aldo Leopold, in the central essay of *A Sand County Almanac*, examined the role of ecological understanding in creating a permanent, mutually beneficial relationship between land and people. That essay, "The Land Ethic," is widely recognized today as the distillation of Leopold's environmental thought. In it he urged the adoption of a new mode of thinking about the human relationship with the land: a land ethic. The evolution of that land ethic, reflecting concerns of human use and ecology, would lead to a new form of stewardship for the land. Leopold famously contrasted the land ethic with an older way of thinking about the land, describing the split as a "cleavage" of a new form of thought. As the older, more established group, which he terms Group A, "regards the land as soil, and its function as commodity-production," a newer, still evolving Group B "regards the land as biota, and its function as something broader." Leopold described the evolution of ethics toward both the ownership of land and its proper use.[2]

At the heart of that essay, Leopold paused to look back on his own career, offering a summation of early-twentieth-century American forestry distilled into a few bitter sentences. Leopold turned to the discipline of forestry, the discipline in which he had been trained, in which he had worked for many years, and which he still claimed as "my own field." He used forestry to illustrate the possibility of evolving a land ethic, while not quite acknowledging that the field itself may have been the motive force behind his articulation of it. He described Group A's way of thinking about forests as "quite content to grow trees like cabbages, with cellulose as the basic forest commodity. It feels no inhibition against violence; its ideology is agronomic." However, he detected the inklings of a Group B evolving within forestry and described this new way of thinking as "prefer[ring] natural reproduction on principle. It worries on biotic as well as economic grounds about the loss of species . . . [and] about a whole series of secondary forest functions: wildlife, recreation, watersheds, wilderness areas. To my mind, Group B feels the stirrings of an ecological conscience."[3] Leopold's message is clear: Group B shows a newly emerging sophistication within forestry and is taking the first step toward the expression of the land ethic.

The historical sketch of forestry Leopold drew in "The A-B Cleavage" is not accurate, of course. He filtered and simplified his own history as part of the rhetorical strategy in the essay. Indeed, the great value of *A Sand County Almanac*, the root of its enduring appeal, is in the persuasive power Leopold develops through the chapters as he builds a deeply appealing vision of an attainable ecological holism. He wrote the book as an environmentalist, not as a historian or as a memoirist.[4] He aimed to show land management splitting into two discrete entities with incompatible worldviews, which abetted his main aims. But in doing so, he also reshaped history as a reflection of his own professional experience through several tumultuous decades. The quick and clean "cleavage" Leopold depicted was, in reality, a messy, contentious, and incomplete separation. Leopold's characterization of forestry in *Sand County* reflected not only the divisions within the field but his own cynicism with it. Moreover, although in The A-B Cleavage he assumed the position of an interested bystander, in truth he had been promoting the alternative view, the "ecological conscience" of the field. The A-B Cleavage, as Leopold referred to it, was the fragmentation of a profession, but in the process of that fragmentation both the A and the B Groups were changed. Leopold heralded the B Group as pioneers of a new form of environmental thought and action. However, the A Group, left holding the reins of professional forestry, was also affected when the B Group split off. *Money Trees* examines this split, both within the ranks of professional forestry and within some of the forests those foresters managed.

As *A Sand County Almanac* has grown in readership and influence since publication, Leopold's bitter characterization of the profession of forestry has often been accepted as accurate. Indeed, the rhetoric of mainstream environmentalism has often incorporated a tacit assumption that professional foresters are intransigently conservative and anti-ecological. This has led to widespread misunderstanding of both the history of professional forestry and the history of twentieth-century forests. To understand historical change in the profession of forestry, and by extension change in American forests themselves, we must acknowledge that Leopold's tale is not the whole story. Examining the A Group's approach to the forest is just as important as examining that of the B Group. The changes that mainstream forestry undergoes as it experiences this internal dissension affect the trajectory of American forest practice and policy for decades to come. This book follows American forestry from the beginning of the twentieth century to the 1940s. I examine the shifting allegiances and dissensions among the multiple stakeholders in these forests: scientifically trained foresters, the lumber industry, and the political and popular forces advocating for resource conservation. By following the developments within and between these groups of stakeholders, we can come to understand the real nature of that shift in thinking, which Leopold proclaimed as the "first inklings of an ecological conscience."

That essay in *A Sand County Almanac* reflected Leopold's frustration with the direction the profession had taken over the years he had known it. A teenage Leopold had decided on a career in forestry in 1900, entranced by the promises offered by the newly created Yale School of Forestry. He would, in a sense, watch the profession grow up with him. As he worked in a progression of governmental, private, and academic posts he found himself increasingly dismayed at the path it had taken. The harsh critique he offered, the unbridgeable chasm he depicted between Group A and Group B, echoed with personal regret. He was writing about a real rift in the philosophy of land management, between mainstream forestry and the radical visions embodied in wilderness activism and ecological management. However, his depiction of the historical cleavage, stirring though it is, does not tell the whole story of the complicated relationship between the mainstream and its discontents. We can respect Leopold's poetic passion in *A Sand County Almanac*, but we should also examine the real history of how forestry fractured, and why. Behind his caricature is a layered tale that, on both sides, may be far weightier than the simple cleavage he portrayed.

The immense power of American industrial capitalism to form agendas for resource management was a driving force for change in the profession. "The Land Ethic," despite its strength as a call to action, did not address the role of industry in shaping American landscapes. However, the economic concerns of the lumber industry played a central role in shaping the professional concerns of American foresters. Changes in professional forestry were spurred by shifts on the ground level, on the front lines of lumber extraction. And in the early twentieth century, the most intense area of development in the logging industry was the Douglas fir forests of the Pacific Northwest. The political urgency and economic pressures of the fast-evolving Douglas fir lumber industry spurred introspection and infighting within the profession of forestry. The technology for extracting lumber from rugged landscapes improved dramatically, to the point that loggers could quickly and easily clear sites that would have previously demanded long and painstaking effort. Faster logging methods were matched by streamlined processing and distribution networks, amplifying both the economic pace and the ecological impact of the American lumber industry. The skyrocketing efficiency of lumber extraction was a shock to the status quo that caused a number of prominent foresters to question the fundamental intellectual assumptions of their profession. The economic pressures of industrial forestry grew as the impact of logging increased. These Douglas fir forests had become the nation's most important site of lumber production by the 1940s. This rise deeply affected foresters' sense of the capabilities of the lumber industry, and more than that, transformed notions of the value of forests.

The History of Forests and the History of Forestry

Foresters had a complex relationship with the forests that were the basis of their profession. Overall, education and research in American professional forestry were dominated by an agenda set by those in the upper echelons of federal forest management. Foresters were scientifically grounded but were often also in government employ, managing public land and setting policy. The effect was that a great number of foresters worked at some remove from their subject of study. Indeed, the profession was centered not in the forests that were its focus, nor in the halls of academia, but instead in Washington, DC. The rapid growth of the Douglas fir industry was managed from a distance. Those on the ground in Douglas fir country were directed largely by decisions made in Washington. The national and, increasingly, global reach of lumber companies meant private foresters mirrored public foresters' delocalized approach to management. The economic issues for the lumber markets of the Northwest could not be untethered from national economic factors, or from the framework of corporate taxation. Foresters therefore needed to maintain a wide perspective even when they worked in specific places. While the evolution of American forestry resembled, in some ways, the professionalization of any number of other scientific disciplines, it differed in that its focus was often extremely site specific. Problems in forestry proved difficult to abstract or generalize. Foresters, unlike many sorts of scientific professionals, had to continually respond to developing problems as industry acted in very particular locations and landscapes. As a result, the science of forestry, despite the Washington-centered hub-and-spoke nature of its employment patterns, could never be as untethered from place as other scientific disciplines.

Forestry is not a narrow discipline. Its aims are not only to study the forest but also to accommodate economic use of the trees in that forest. Early-twentieth-century foresters sought to define their profession as scientific but simultaneously interacted with lumbermen, politicians, and other nonscientists. The lumber industry subsists on forest products, and as the supply of virgin forest dwindled lumbermen became increasingly invested in the continued productivity of those forests. Maintaining the longevity and productivity of those forests has always been important for the economic well-being of both local communities and large corporations, as well as for the ecological health of the biota therein. While there were—and still are—intense disagreements over how best to achieve this goal, the biological realities of the forests must always be acknowledged as the limits of economic possibility. Defining satisfactory ecological and economic parameters of a particular forest type is a reflection of the values and politics of those who keep them. In the decades under scrutiny in this book, the problems of the Douglas fir forest often embodied the problems of the nationwide lumber industry. At the same time, national concerns often played out through debates over local Northwest issues. Thus, to understand forestry historically, it is necessary to follow the story on all the levels that foresters were engaging with historically: both regional and national developments, both biological and economic considerations.

The business of lumber has shaped American forestry. Professional foresters work not just within an intellectual community of scientists and government employees but also within the realm of forest owners, lumber industry managers, and loggers. While federal employees dominated much of this history, the conduits of intellectual exchange between federal foresters and others were important to the development of the profession. While the government often set land management agendas that others were compelled to follow, industrial influence grew steadily as logging in the Northwest increased. By the 1940s, many professional foresters moved easily between federal employ, academic posts, and industry consultancy over the course of their careers. It is impossible to isolate the research side of forestry from the management side, and one cannot understand the development of forestry without examining both scientific and economic influences on the profession. This book, then, shows the science and management of American forests increasingly affected by the mainstream American forest industry, eventually resulting in a rift within the profession. It follows foresters like Leopold who, alienated by the increasing subservience of their profession to the goals of resource capitalism, found alternative professional paths toward land management.[5] In addition, I show how the lumber industry, responding to the increasing prominence of professional foresters, responded by reformulating existing forest science to serve its own purposes. The pace and scale of the industrial use of the Douglas fir was a constant background to the evolution of the profession and its discontents. Fears of mismanaging the seemingly irreversible consumption of these pristine stands of giant trees haunted the profession. Those foresters who objected to the increasingly industry-oriented stance of their discipline responded with radical critiques, creating a current of dissent within the professional community. Worries about misusing the great, yet dwindling, stands of untouched forest became a common complaint, as numerous parties claimed primacy of agenda and method.

As a book tracing the development of the profession of American forestry, *Money Trees* owes much to the methodology of the history of science. Foresters saw themselves as scientists, and it is useful to examine the path of their discipline in light of developments in other American sciences, especially ecology. The authority foresters asserted for forest management prescriptions was based in scientific research into the biology of forests and the economics of the United States. The history of forestry, especially forest biology, can thus be taken as part of a larger story of American science in the first half of the twentieth century. Ecological approaches to field biology grew in prominence early in the twentieth century, although researchers did not always identify their work specifically as ecological. Examining the influx of ecological thinking into forestry research is a major project of this book and requires historical examination of the practice of field ecology for a full understanding.[6] Ecology was, during this time, a descriptive science, focusing on defining and characterizing the interactions and interdependence of species and their environments. Conversely, forestry was a prescriptive science, uniquely subject to the needs and goals of federal land management and industrial resource exploitation. Thus, for the historian, American forestry cannot be viewed just within the context of the history of biology but must instead be seen as enmeshed in a larger fabric of American political and economic processes.[7] The evolution of forestry does not follow the classic path of disciplinary professionalization. This is not a story of the simple triumph of one research agenda over others, or of the resolution of conflict through refocusing professional identity. Instead, this book offers a more complex tale. The practice of American forestry is a story of the twining and untwining of that discipline with other human pursuits: ecological research, wilderness activism, land management, and timber harvest. All these pursuits, practiced under the looming specter of twentieth-century resource capitalism, attempted to impose order on nature in some way. The intellectual closeness of these endeavors to one another affected the way all of them would develop.

While the demands and pressures of the regional Douglas fir industry are at the forefront of this story, the book's focus is on the professionals shaping the field of American forestry. The efforts to control and manage forests and other public lands have often been topics for environmental historians. National policies and politics have had tremendous impact on the American environment and on patterns of resource use. The US Forest Service has been a central organization for management and policy regarding public lands.[8] Studies of the federal control and management of forests have done much to uncover the historical connections between economic needs and ecological realities, and the often unforeseen results of such management. Some environmental historians focusing on forest management of particular regions or locales have focused on the local story at the expense of the national context. Some scholars have mistakenly neglected foresters employed outside the federal agencies in industrial, activist, and academic contexts. *Money Trees* expands the historical scope of the subject to encompass a wider array of foresters, with the goal of examining forestry's critiques and discontents alongside the formation of federal management policies.[9] The fine-grained environmental changes that occurred within these forests are the object of their efforts and the focus of their goals. These forests, then, are the objects of foresters' intellectual work, and the environment's responses to forest management are the central focus of their decisions.

This book begins at the turn of the twentieth century with American forestry's birth as an independent, scientifically based profession; follows foresters through decades of challenges in understanding and managing the American forest; and ends as the discipline, impacted by the fallout of the New Deal and the Second World War, fractures in the mid-1940s. The complex dynamics of this history compel the inclusion of voices from the lumber industry, from federal foresters, and from the critics of both. A small group of prominent foresters maintained a strong voice in these debates from the start and were responsible for much of the introspection and debate over what forestry should—and should not— do for the nation and the forest. Led early on by Gifford Pinchot and Raphael Zon, this group was powerful in Washington and yet ideologically radical. That coterie of foresters attracted Bob Marshall, and eventually Aldo Leopold, with a vision of forest management that was not beholden to industrial pressures or political exigency. The dissolution of this coterie coincided with larger shifts in the professional landscape as the Second World War loomed. What Leopold bemoaned as the "cleavage" came about with the rise of a production-driven view of forests, adopted by most in the mainstream of the industry. It resulted both in the industry's co-optation of simplified forest science for private forests, and in the federal adoption of sustained yield as a management framework for public forests.

The first two chapters examine the roots of the transformation of the Douglas fir lumber industry into a scientific, highly mechanized entity crucial to regional economic health. Chapter 1 examines how the forest industry turned to the Pacific Northwest at the beginning of the twentieth century, thus establishing the Douglas fir forests as the setting for the events to come. Years of logging in Michigan, Wisconsin, and Minnesota had provided lessons in how best to extract value from timberland. The boom-and-bust cycles of the market had also fueled the consolidation of the myriad lumber operators into an industry dominated by a handful of large companies. As the forests of the upper Midwest became increasingly exhausted, these corporations looked toward fresher lands in the West. This westward movement was the result of agreements and transactions between some of the wealthiest industrialists in the nation. Frederick Weyerhaeuser led this movement, with the purchase of a massive swath of untouched Douglas fir forest from the railroad magnate James J. Hill. In the succeeding years, other lumber giants followed Weyerhaeuser's lead. This expansion affected how these corporations understood the economic possibilities and ecological constraints of the Douglas fir forest region. The scientific understanding of the forest was in a state of change during this period. This chapter shows how the shifts in biological understanding of complex ecological entities affected industrial approaches to exploiting the economic potential of the forest.

Chapter 2 explores the federal reaction to the consolidation and westward movement in the lumber industry, through the creation of the US Forest Service. The founding years of the Forest Service have been well documented. Many are unaware, however, of the innovative intellectual underpinnings of the science and of the prominence of research in those early years. The individuals responsible for setting the scientific agenda of forestry within the Forest Service in these early years were significantly influenced by the new ecological approaches then gaining prominence in plant biology. Forest research was influenced by new ecological modes of thinking in early-twentieth-century American forestry. Early Forest Service leaders thought developing a program of federal field research was important. Raphael Zon and Gifford Pinchot were both influenced by ecological approaches to research and management as they shaped the federal plan for forest research in the early years of the Forest Service.

The third chapter examines the effects of that ecological approach to forest science on federal foresters in the late 1920s. Such an ecological approach offered not just different methodological approaches to forest science but also a different way of understanding the role of scientists in forest policy decision making. This chapter examines how some foresters made use of ecological innovations to understand the forest in a deeper way than before. Studying the forest as a whole, a site of complex interactions and long-term developments, was made more reliable by viewing the forest through an ecological lens. Thinking ecologically about American forests led to new avenues of scientific understanding but also posed challenges for the integration of scientific findings into the framework of federal forest policy. The research aims of ecological science often seemed at odds with the goals of the US Forest Service. Some foresters worked to find a way to fuse ecological forest science with the sense of social utility that had come to define American forestry. This chapter examines the early career of Bob Marshall, who linked his interests in forest recreation and wilderness advocacy with a continuous program of ecological research. A close allegiance grew between Zon, Marshall, Pinchot, and a small handful of others as they sought a way to reinvigorate the social conscience of the American forester. Their conception of nature, and thus the root of their arguments for wilderness preservation, grew from an allegiance to the modern science of ecology. A full understanding of the particular ecological view held by Marshall, as well as the relationship between ecology and politics, leads to new insight into the roots of the modern wilderness movement. As Marshall and others drew closer to the ecological mode of inquiry, they broke away from the clearly articulated economic utility of mainstream forestry. Marshall's sense of himself as a scientist and as an ecologist, which was incubated during these years, inspired his conception of a new sort of social utility for forestry. A coterie of like-minded foresters alienated from the mainstream of their profession came to formulate a strong activist agenda, which included methods and goals for forestry that resulted in turmoil and eventually created deep rifts within the profession.

Chapter 4 discusses the changes in federal forestry in relation to the dramatic changes in the profession seen during the New Deal years. Economic uncertainty, coupled with changes in federal land and resource management, caused turmoil for foresters. This was a time of introspection and disagreement for foresters as they tried to define their profession in the face of great changes both in the lumber industry and in the management of American public lands. The Zon-Pinchot coterie wanted both to make forest research more rigorously ecological and to steer the profession in a less industrially oriented direction. While adopting an ecological approach could enrich and sustain forestry's scientific value, it also had drawbacks. Thinking of forest research as ecology lessened the traditional social utility that was an indispensable attribute of the profession, an unexpected result for ecologically minded foresters. With the new pressures and opportunities presented by the New Deal, the coterie's critiques changed. In a time of government upheaval they saw a chance to remake forestry as an ecologically informed and socially oriented profession. They demanded that the profession maintain distance from the lumber industry and steer the agenda for forest research away from focusing on the problems of yield and production that were a priority for the industry. It became clear, however, that the majority of foresters were reluctant to make such a change. Meeting the goals of the lumber industry became increasingly central to the profession of forestry. Alienated by this turn of the profession toward industry, the coterie of radical foresters sought a different way to marry social utility to forest research. These foresters became the core of the newly formed The Wilderness Society. Land-use activism became a way to keep industry at arm's length while retaining a professional focus on the public good.

Chapter 5 explores the changing relationship between professional foresters and the Pacific Northwest lumber company executives. The largest companies in the American lumber industry, focused on the dwindling acreage of untouched Douglas fir, fought against increasing public criticism and threats of regulation. The New Deal amplified federal commitment to impose order on all sorts of industrial land use, including logging. Faith in the possibilities of scientific forest management drove federal and state foresters toward advocating sustained yield as a management framework for public lands. Concerns about destructive and careless forest use by large corporations led many politicians to advocate regulation of logging on private lands as well as public. The largest of such companies, the giant Weyerhaeuser conglomerate, worked to counteract the growing criticism with a public relations strategy of its own. Rather than submit to the regulatory strictures of sustained yield, it recast itself as the steward of an agricultural crop. If timber was a crop, it argued, then lumber companies should be trusted to farm it indefinitely. This chapter shows how the ecological strain of forest research, which had been understood as so integral to the Forest Service in earlier years, suddenly became relegated to a minor role in forestry. The final blow for the cohesion of the Zon coterie of radical foresters came in late 1939, with the untimely deaths of two members of the group: activist forester Bob Marshall and Forest Service chief Ferdinand Silcox.

The final chapter concerns professional forestry's turn toward industrial lumber production during the early 1940s. Foresters were called on to manage forests using a more sedate, long-term approach to harvesting, as companies eschewed the old cut-out-and-get-out logging. Both industry and the federal government introduced schemes for forest management that distanced professional forestry from its origins in natural, wild forests. In 1941, Weyerhaeuser Timber Company established the Clemons Tree Farm on its landholdings near Montesano, Washington, where it still exists. Spearheaded by the company's first dedicated public relations executive, the Clemons was a corporate ploy for public approval as much as a privately funded method for second-growth lumber production. Shortly after, the sustained-yield forest management pioneered in the late 1930s became a national goal. Foresters in both public and corporate settings developed an increasingly industrial approach to the forest.

Money Trees demonstrates the linkages between ecology and policy. The Douglas fir forests promised great fortune for those who could log there, but those potential riches were paid for by sacrificing the original untouched forest. The Douglas fir lumber industry became so lucrative in part because of the specific ecological attributes of the tree, the forest, and the region. However, the industry's success can also be attributed in part to the relationship between on-the-ground realities of the forest and the national development of forest science and forest management. As the great old-growth Douglas fir trees were felled, foresters scrutinized the natural forest, experimented on the clear-cuts, controlled logging rates, and debated management decisions. Trees growing in the coastal Northwest were of interest not only to loggers and locals but also to the policymakers in Washington, DC. This book examines how each side of this relationship affected the other. The unique ecological and economic realities of the Douglas fir forest affected the development of a federal forest agenda, as well as the collective professional goals of forestry. Likewise, professional foresters shaped the Douglas fir forest, not just of their time but for generations to come, through their research and policy decisions. This book follows the transformation of the Douglas fir forest from a natural forest into an industrialized site of lumber production, and it explores how foresters helped, hindered, and commented on each stage of that transformation.

The American Northwest's wild Douglas fir forests teem with life. Although this forest type's common name is derived from its most iconic species, numerous other tree species thrive alongside the Douglas fir and compete with it for resources. Insects, herbaceous plants, mosses, birds, and animals are also all vital parts of the forest. Streams, weather, terrain, fires, even people are all encompassed within the forest. The forest, we tell ourselves, cannot be ecologically understood as simply stands of Douglas fir. We cannot focus only on the trees—while ignoring the understory, the wildlife, or the soils—to understand the forest. The first generations of American foresters, however, often did something very close to that. The Douglas fir was money—money in tree form. *Money Trees* examines some of the many people who saw the forest only for that extraordinarily lucrative timber tree. Lumber companies considered the forest either for how it affected extraction of Douglas fir logs or for how it influenced living Douglas fir, but not for its own sake. The work of early professional foresters in the region was often simply a matter of determining ways to log, protect, or manage the Douglas fir, that money tree. When foresters did begin to incorporate an ecological approach to the forest, they implicitly attacked the primacy of economic considerations. By denying that the money tree was the only important thing in the forest, ecological foresters rocked their profession. By allowing the money tree to rule management decisions, industrial foresters transformed Northwest landscapes. The value of Douglas fir, and its place within the forest, were issues at the heart of this fight.

The early days of commercial Douglas fir logging in the rich forests along the Lower Columbia River (c. 1905). Courtesy Oregon Historical Society.

1 Northwest Promise
New Forests for a New Century

At the dawn of the twentieth century, the dark and timeless forests of the Northwest coast were suddenly thrust into a bright new world of modern industrial capitalism. On January 3, 1900, Frederick Weyerhaeuser purchased 900,000 acres, or 1,406 square miles, of western Washington forestlands from railroad magnate James J. Hill. With this one massive transaction the midwestern lumber baron became the major player in the Northwest lumber industry. The deal had been struck after long negotiation in the drawing rooms at 240 and 266 Summit Avenue in Saint Paul, Minnesota, the neighboring mansions of these two titans of industry. A breathless press reported that Weyerhaeuser had put up the almost unfathomable sum of $3 million dollars that day, with an additional $2.4 million to be remitted through eight semiannual payments. The value of this purchase was not in the land itself, much of which was rugged to the point of inaccessibility. No, it was what stood upon it that merited the price: the last forest in the nation yet to see significant logging. And what a forest it was. Unlike any other, it was dominated by massive Douglas fir growing so tall and dense that they threatened to blot out the light from the sky.[1]

A month later, as George S. Long stood on the forest floor below these giants, the forest betrayed no evidence of its transformation into capital. Long, dispatched by Weyerhaeuser from Saint Paul to oversee the new lands, was astounded at their extent and their wildness. This was nothing like the lumbering he knew from Minnesota, where well-mapped lands gave up their timber without protest. "When I came out to the Coast," he recalled later, "[the] business was solely and strictly timber lands, and what I did not know about that would fill a book." The management expertise that had pushed him up the ladder of lumber management had come behind a desk, and he "had never been in the woods in my life." He had no knowledge of these trees, or how to determine "why one tree might be worth more than another, and absolutely knew nothing about logging; and here I was confronted with a job where I was expected to tell others what to do!"[2] The midwestern pine forests had yielded manageably sized trees, which oxen teams dragged from gently rolling hills to float to mill on the navigable rivers crisscrossing the landscape. Weyerhaeuser had grown rich on those trees. But the Douglas fir, rough giants clustered on ragged mountains, would not give up their value so easily.

Not only were these new lands unmapped and little understood, but the Douglas fir, the money tree of this new land, was nothing like the trees of the Midwest. These forestlands held immense amounts of high-quality timber, but accessing that wealth would prove difficult. The forests of the Cascade and coastal areas of the Northwest differ greatly from those of other regions of the country. While these forests experience relatively mild and rainy winters, they have very rugged, steep terrain. The trees in these forests are not naturally a mix of species but are dominated by a single species, the coast Douglas fir, along with undergrowth such as fern, manzanita, and rhododendron. The Douglas fir is the second-tallest tree species in the world after the coast redwood, regularly reaching 250 feet in height and 6 or more feet in diameter at maturity. The trees often take five hundred years of slow growth to reach this mature size—and individual trees have been measured at one thousand years old. Because the trees are so large and grow relatively densely, the yield of lumber per acre in the region's old-growth forests was higher than anywhere else in the country.[3]

Before the United States had gained control of these lands, the Hudson's Bay Company had overseen an increasingly lucrative timber trade. As the British influence ebbed, American lumber operators moved in with increasingly ambitious plans. Logging and milling operations along the shores of the Northwest shipped lumber to the growing cities of the California coast. They also found markets for Douglas fir much farther away: meeting the building needs of the increasingly deforested islands of Hawaii, shipping regularly to the boomtowns of Australia, even supplying a shipload of fifty-foot-long timbers for Cecil Rhodes's stately home in Cape Town, South Africa.[4] Stymied by the difficulties of logging in the interior, for the most part they could harvest only from the river mouths and tidewater edges of the great forest. A great wave of settlers came into the Northwest, encouraged by the government to establish homesteads. The most promising agricultural lands were those of the wide, flat floodplains of the Willamette River, between the Coast Range and the Cascade foothills south of Portland. These lands, and other scattered fertile and flat areas, were soon developed for agricultural use. The settlers soon found, however, that many of the homesteading practices of clearing forestland for fields, which had so significantly abetted the development of the eastern parts of the continent, were simply not possible here. Douglas fir trees were immensely larger than the trees of the mixed hardwood forests of the East, dwarfing the humans who tried to fell them. At its base, the Douglas fir's bark was so fibrous and gummy that it was almost impossible to break through. To cut clean through the trunk of a mature tree was simply beyond the capacity of the settlers' equipment. Even the simple procedure of removing a stump promised to become a herculean task in these forests. Attempts to harvest by other means, including lighting fires within holes bored in the trunks, were unpredictable and slow.[5] Despite the settlers' ingenuity, facing such challenges before any money making could begin made the Douglas fir forests unappealing for individual landownership in the late nineteenth century.

The establishment of small logging companies gradually increased the scope of lumber efforts in the region. As the availability of logs directly along navigable rivers and shorelines lessened, loggers began transporting logs longer distances by oxen teams and by river flotation. They developed small rail lines built out from mill towns into the surrounding forests. The development of the steam donkey and the circular saw revolutionized the industry. The steam donkey supplanted oxen labor with a method for logs to be dragged quickly from the place of felling via a cable and pulley system. The circular saw improved the capacity and precision of lumber mills immensely, which allowed the production of high-quality lumber at lower prices. In the mild and damp climate of the Pacific Northwest, Douglas fir trees regularly grow to heights of 175–200 feet, almost always with a perfectly straight, nonbranching trunk. Hence logging an acre of Douglas fir will yield far more board feet of lumber than will logging an acre of most other species.[6] Furthermore, the lumber produced from the Douglas fir is dense and straight grained, making it ideal for house frames, floorboards, and veneers, some of the most lucrative wood markets. Some of the other tree species in the region, such as western red cedar and western hemlock, were not nearly as prized by lumber companies, who often preferred to leave those trees standing on a site than make the effort to cut them. As western lumber manufacturers began publicizing the value of the Douglas fir over other, more familiar softwoods, the demand for it grew exponentially as builders and lumberyards became familiar with its versatility.

Although the completion of the transcontinental Northern Pacific Railway into the Puget Sound area held the promise of new markets, rail freight fees remained so high that for a time it was still cheaper to ship lumber across the Pacific to Australia than overland to the East Coast of the United States.[7] When the Great Northern line was completed, its president, James J. Hill, announced that his freight fee would be a third less than the Northern Pacific fee. The Northern Pacific immediately matched the new lower fee, and with this the opportunities for selling to the cities east of the Rockies became more feasible. James J. Hill lowered the freight rate because he understood that the revenue from transportation fees was only one of several important contributions to a railroad's bottom line. After hauling settlers and goods out to the Northwest, the train cars would need to return east, full or not. Loading them with timber at a cut rate meant the railroad was not moving empty cars and was still generating some money through the fees on the timber. Hill had built his own Great Northern without government land grants, but he came into possession of grant lands when he took over as the president of the Northern Pacific Railway. While the federal government had intended the land grants to be sold to individual investors and small landowners, there were few buyers, especially in the Northwest. The assets of the railroad grant lands of western Washington were mostly undocumented, but from what had been seen during the process of surveying and building it was clear that large quantities of immense Douglas fir covered the hills and mountains. Selling parcels of forestland to individuals was a laborious and fickle process, and those individuals might never generate any freight for Hill to move east. Profit could be much simpler through the Weyerhaeuser deal, and Hill could generate the funds he needed immediately. Further, by dealing with Frederick Weyerhaeuser, Hill could ascertain that he would soon have a friendly logging company generating timber freight for his trains to move eastward to market.[8]

Hill's deal with Weyerhaeuser would benefit Weyerhaeuser as well. The Midwest's lumber boom years were coming to an end as the output of the increasingly devastated forests of the northern Great Lakes ebbed away. But even for a successful concern like Weyerhaeuser, finding that sort of money was difficult. Weyerhaeuser, an aggressively growing company, had become one of the major lumber companies of the Midwest through buying dozens of smaller companies in the region every year. The Weyerhaeuser name was undoubtedly one of the best known in the lumber and logging industry. The founder of this lumber empire had been Frederick Weyerhaeuser Sr., a German who had immigrated to the United States in 1852, where he found work in a lumberyard in Rock Island, Illinois, on the banks of the Mississippi River. By 1860, Weyerhaeuser and his brother-in-law Frederick C. A. Denkmann had started a sawmill of their own in Rock Island. The business partners set their sights on acquiring forestland in northern Minnesota and Wisconsin. Soon, the firm of Weyerhaeuser and Denkmann was managing large timber operations in the white pine forests of northern Wisconsin and Minnesota, and Frederick Weyerhaeuser moved his family to Saint Paul, Minnesota, to be closer to his work. By 1875 lumbermen throughout the upper Midwest recognized Frederick Weyerhaeuser as a leader in the industry, as he rapidly expanded the Weyerhaeuser holdings. Of his seven children, all of the sons entered the family business, as would many of his grandsons and great-grandsons. The Weyerhaeuser family became known for buying out struggling lumber companies and incorporating their holdings as subsidiary companies.[9] The steadily expanding collection of linked corporations and subsidiary companies that existed under the Weyerhaeuser umbrella was one of the largest and most financially stable conglomerates in the nation.[10]

Upon purchase, Weyerhaeuser and his associates had only sketchy knowledge about what they had on their hands. Raising the capital had meant not only marshaling all the available funds of Weyerhaeuser's Minnesota lumber operations but also partnering with other investors to raise the full amount. Industry observers speculated that the lands might not even contain much timber, and that the decision to buy might have doomed the company's financial future. But after his initial survey Long wrote Weyerhaeuser that the timber was so thick that a man could barely walk through the forest to inspect it, and the trees were taller than any in the Midwest. The potential profits for Weyerhaeuser were enormous, once logging operations and log transport could be established in such inaccessible locations. But logging was difficult, as the area was poorly mapped and not yet crisscrossed with roads. Further, few had real expertise in logging these areas. Long resorted to hiring a railroad engineer to plot a feasible rail spur through the rugged terrain without steep grades or sharp turns, at the same time hiring a "practical logger" with experience building roads for logging, and he then simply picked between the two: "when we get these two opinions as to where the road should go, I think we can then decide which is the most feasible place to put in the track."[11] Long bought lands that bordered those Weyerhaeuser already owned, both increasing the company's financial investment and making logging access easier. Once the logs were harvested, however, the region's sawmills were also sorely lacking. Long reflected to his employer that it was "very apparent that we will soon have to drift into the manufacturing of this timber [ourselves]," and that if so, "it would seem better for the W. T. Co. to own all of this property if there was a probability of the timber being manufactured."[12] Frederick Weyerhaeuser agreed with Long, and the company began to establish processing operations in the Northwest. Weyerhaeuser both built new sawmills and increased the capacity of existing ones. As the region's log transport and manufacturing infrastructure developed, many of Weyerhaeuser's competitors also began to expand operations into the Northwest.[13]

Early-twentieth-century federal foresters made management decisions largely in agency offices in Washington, DC, but the dramatic changes in the Pacific Northwest lumber industry were never far from their minds. The logging operations that moved into the region during these years were already fully developed industrial corporations working on a national level. Following the Northern Pacific–Weyerhaeuser land transfer, myriad other lumber companies similarly moved focus from the Midwest to the Northwest. The power of these giant corporations in a poor and lightly settled part of the country seemed unimpeded. Federal land managers wished to keep the vast resources of the far west under federal control, even when they stood on private land. They believed regulation and tax structures should control who could access the lumber wealth, as well as when and how. Many argued that state and local corruption in Washington and Oregon politics would lead to an overbearing influence of the lumber giants. Regional politicians favoring industry could make it simple to quickly strip the land of its timber wealth and leave untold wastelands behind. Midwestern experiences with destructive wildfires had proven that the results of careless logging could be devastating for communities. Much of this forest was on steep slopes, and logging destabilized the soils. Mudslides during the rainy season could be severe. Watershed contamination and the effects of silting on fisheries also loomed large for the rugged Northwest. For the lumber industry, the effect was to tie its regional activities to the politics and policies of the federal government. As it generated increasingly substantial lumber profits, the specific management issues of the Douglas fir forests grew in importance to the agenda of federal foresters.[14]

Forest Science for a New Century

Many historians have noted the increasing reliance of the Progressive Era government on scientific expertise in various avenues of planning and decision making. American professional forestry was part of this overall trend, as the rationalization of forest use required developing a framework of scientific knowledge on which to depend. Indeed, this was a time when turning to scientific content was seen as the way to be sure that the plans would be worthwhile. However, there has been comparatively little historical examination of the texture and character of the science that contributed to the construction of this profession. Not everything that falls under the umbrella term "science" is the same. Pinchot and other Progressive Era foresters chose to employ a particular ecological mode of scientific inquiry that influenced their basic understanding of the forest. Upon this ecological foundation, a template for understanding and managing forests was built up. An understanding of the contributions of this new forest science will shed new light on the management decisions that foresters made.

In the early twentieth century, many of the botanists interested in the new field of ecology entertained ideas of establishing rules to explain the interaction of species in a particular location. In the late nineteenth century, European plant geographers first articulated the vegetational climax as the natural state of the vegetation at a site, determined by climate and topography. It was extremely stable and was thought to replace itself indefinitely in a given location unless interrupted by fire, landslide, or other catastrophic event. Furthermore, the climax was thought to be composed of the species most ideally adapted to a site, species that had evolved specifically in response to the site's requirements. Thus, a desert, forest, or grassland could be identified by its climax species, with climate and species inextricably linked. The idea of the climax was accompanied by that of succession, the process of progression toward the climax. The rules of succession were similar to those of Ernst Haeckel's evolutionary premise of recapitulation, prevalent at the time. That premise claimed that the development of a single human being from embryo to infant mirrored the evolution of the human species from single-celled organisms, through protoamphibians and early mammals, to the human form. Both of these ideas held weight mainly for their predictive ability and for the way a simple axiom could make comprehensible what had previously been seen as undifferentiated or chaotic phenomena.[15] While there were many individuals involved in the development of the succession-and-climax theory in ecology, two were especially important in the context of forestry. Henry Chandler Cowles and Frederic Clements both addressed the impact of the succession-and-climax theory for applications in forest management and ecology. Also, works by each were widely read by foresters as scientific support for their understanding of the ecological dynamics of plant communities.

Cowles, a plant geographer based at the Hull Botanical Laboratory of the University of Chicago, performed the foundational research for plant ecology in the United States in the mid-1890s. Cowles studied the plant life of the three-mile-wide swath of sand dunes on the southern shores of Lake Michigan, in the area that is now Indiana Dunes State Park. Strong winds off the lake move the dunes around regularly, burying growing plants and constantly changing the topography. A newly formed dune, then, "by burying the past . . . offers to plant life a world for conquest, subject almost entirely to existing physical conditions."[16] The newest dunes and those closest to the lakeshore are mainly sand dotted with Cowles's so-called primitive assemblages of plants, mainly hardy herbs and xerophytic grasses. As one moves farther back from these areas one sees progressively more plant cover, of a progressively more complex sort. Cowles explicitly linked this spatial progress back from the newest dunes with the temporal progression of plant species in a single idealized location. He stated that "in the historical development of a region the primitive plant societies pass rapidly or slowly into others; at first the changes are likely to be rapid, but as the plant assemblage more and more approaches the climax type of the region, the changes become more slow." Likewise, as one moves away from the xerophytic plants of the fresh dunes, the "dunes pass through several stages, finally culminating in a deciduous mesophytic forest, the normal climax type in the lake region."[17] Cowles's work with the sand dunes showed a real-world example of the revegetation of an area left devoid of plants. His description of the orderly progression from bare ground to a self-sustaining climax forest appealed to the desire in many ecologists of the early twentieth century to find rules that governed the overwhelming complexity of ecological observations. His implication that the changes in the plant community of a single site over time mimicked a gradient of increasing ecological complexity across a length of space satisfied as well.

Charles Darwin's immensely widespread writings on evolutionary theory also spread the understanding of forest as an interlocking set of organisms, factors, and phenomena. Darwin, in the process of explicating his theory of evolution by natural selection, laid out an early model of ecological thinking for scientists. To explain the importance of competitive interactions between species, Darwin offered a view of the natural world that focused on the interrelations of the species with each other and with the environment in which they lived. In Europe, where much of the landscape had seen centuries of human use, there were few examples of these struggles still in stark effect. Darwin turned to other parts of the world, including the United States, to provide a powerful vision of the struggles for existence. He wrote that

> the trees now growing on the ancient Indian mounds, in the Southern United States, display the same beautiful diversity and proportion of kinds as in the surrounding virgin forests. What a struggle between the several kinds of trees must here have gone on during long centuries, each annually scattering its seeds by the thousand; what war between insect and insect—between insects, snails, and other animals with birds and beasts of prey—all striving to increase, and all feeding on each other or on the trees or their seeds and seedlings, or on the other plants which first clothed the ground and thus checked the growth of the trees![18]

Darwin emphasized the dynamic interactions between species, rather than the seemingly eternally stable communities that often greet a visitor to a forest. Only through studying and understanding those interactions could scientists discern the workings of natural selection.

Geographically inclined botanists working in the late nineteenth century approached the study of a forest from an alternative perspective. By trying to comprehend the full array of species at a locality, one could come to understand the differences between such species communities in different locations. Based in the traditions of natural history, American plant geography grew significantly through western land surveys and exploratory botanizing financed by the federal government in the mid-nineteenth century.[19] By the 1890s the potential for new understanding of the variations between plant communities was realized through the work of innumerable researchers in the United States. Charles E. Bessey, at the University of Nebraska, led exhaustive research into the plant communities of the Great Plains, in what would be called the "Founding School of American Plant Ecology." The science of plant geography required thinking of plant communities as relatively static and fixed in place, in order to more effectively link the geographical and biological elements of the study. Hence, plant geographers tended to steer away from discussions of the struggle for existence, instead emphasizing the harmony of nature and the stability of communities. This form of ecology would gain prominence in the first years of the new century, especially through the work of Bessey's student Frederic E. Clements. Clements offered an appealingly structured view of American environments in guides for ecological practice that were widely read by biologists, foresters, and other scientists.[20]

Ecology, then, was in conflict with itself even at birth. Darwin's focus was on the interactions and exchanges between species, creating what can be termed a *dynamic* ecology. Conversely, plant geographers focused on creating full descriptions of plant communities as relatively static entities, creating what can be termed a *synthetic* ecology. Such synthetic ecology could compare different communities to one another, determining factors in the environment by examining differences both in the array of species and in geographical factors. While studies in dynamic ecology would focus on examining the physiology of plants or the behavior of animals, studies in synthetic ecology would focus on finding patterns in vast amounts of data about the spatial distribution of species. Ecologists used both of these modes of inquiry while building up the field, although the two modes involved very different methods of experimentation and observation. While some ecologists endeavored to investigate the structure of plant communities, others looked to understand the interactions within and among populations in a locality. By the end of the nineteenth century, many ecologists recognized the tension between these two approaches to ecology. Indeed, in 1910 the Third International Botanical Congress addressed these inherent differences by formally adopting the terms autecology and synecology. As the field of ecology ascended, its best practitioners and disseminators acknowledged the existence of two separate methodological tracks within it.[21]

Gifford Pinchot was the first American forester to publish a study of forests that engaged with the dynamic, Darwinian ecological framework. Pinchot had attended Yale before traveling to Europe for advanced training in forestry. On his return he worked briefly as a consultant to the Vanderbilt family on the management of their forestlands in the mountains of North Carolina. However, he soon moved on to federal employment, launching a life of government service that is unparalleled in its breadth or impact. Pinchot understood early on that adopting an ecological perspective could help in understanding the forces at work in a forest. In 1899, working in the Division of Forestry, he published *A Primer of Forestry* as both an overview of the field and a statement of American forest practices. His third chapter, entitled "The Life of a Forest," was remarkable for its extensive use of Darwinian terms and imagery. He wrote that the forest was the manifestation of the struggle for existence, and that

> the history of the life of a forest is a story of the help and harm which the trees receive from one another. On one side every tree is engaged in a relentless struggle against its neighbors for light, water, and food, the three things trees need most. On the other side, each tree is constantly working with all its neighbors, even those which stand at some distance, to bring about the best condition of the soil and air for the growth and fighting power of every other tree.[22]

On the one hand, the "community of trees" is interdependent, sharing and creating amenable conditions. But on the other hand, "while this fruitful cooperation exists, there is also present, just as in a village or a city, a vigorous strife for the good things in life," namely light, water, space, and nutritive soil.[23]

Pinchot told a story of succession in the forest that focused on the interactions between trees and the environment, and the competition among the trees for the best access to those life-sustaining necessities. His account of succession was concerned entirely with the actions and interactions at work in the forest rather than with descriptions of the species themselves. Indeed, this account avoided naming any of the species involved in the process of succession. While the trees sometimes depended on each other for mutual aid, such as in sheltering from harsh wind, Pinchot presented such activity as more of a temporary cease-fire than a state of equilibrium. Furthermore, he refrained from describing the final stage of forest growth as a climax. Instead, the trees existed in a dynamic interaction with their environment as they reached senescence and began to decay. He wrote, "As the old trees fall, with intervals, often of many years, between their deaths, young growth of various ages rises to take their place, and when the last of the old forest has vanished there may be differences of a hundred years among the young trees which succeed it."[24] Pinchot's in-depth use of the language and analytical approaches of ecological science and evolutionary theory demonstrates the sophistication of his scientific approach at this stage.

For American foresters, ecological thought opened two different paths into understanding and managing forestlands. Some ecological studies, mainly of sites like grasslands and small lakes, included laborious censuses of the array of species in a location. Such syntheses of data emphasized stability and harmony and meshed with a vision of the forest as a place that was very difficult for loggers or other developers to affect. Synthetic projects like these, when done in forest settings, could underscore the inevitability and robustness of climax forests. Studies that claimed forests were stable and resilient aided prologging arguments, since they could be interpreted as implying that human activity could only temporarily disturb equilibrium. However, other ecological studies emphasized the physiological interactions and constant change within populations. Such ecology focused on the struggle of organisms to thrive in the chaos of nature. Studies emphasizing nature's dynamism in this way consisted of monitoring the behavior and physiology of organisms at a site, offering little in the way of predictive value for foresters. Ecological studies of dynamic change relied more on evidence and experiment than did the theory-driven syntheses that promoted stasis. Such studies appealed to foresters wanting to impact their own field with contemporary scientific values and a modern experimentalist's sensibilities. Researchers used both modes of ecological thinking extensively in early-twentieth-century forest research, although most often without labeling them as such. As we shall see in the next chapter, debates within ecology, and the interactions between forestry and ecology, will have a significant impact on the development of forestry both as science and as management.[25]

Pinchot felt a positive shift in American interest in forestry between 1898 and 1905, coincident with the federal creation of the US Forest Service and the shift of public forestlands to the control of that new bureau. Years later, in his autobiography, he reminisced that "in 1898 the people in general knew little and cared less about Forestry, and regarded the forest, like all other natural resources, as inexhaustible. In 1905 the share of the forest in the life of the Nation was almost everywhere recognized."[26] And this was as true of those men who controlled the nation's industry as it was of the rest of the citizenry. "New recruits" James J. Hill, president of the Great Northern Railway, and Howard Elliott, president of the Northern Pacific, "recognized that the future prosperity of the West lay in the wise use of land and forest." Moreover, "the change in attitude of the lumber industry . . . was even more impressive. It was summed up by the statement of F. E. Weyerhaeuser, son of the head of the great Weyerhaeuser Timber Company, that 'Practical Forestry ought to be of more interest and importance to lumbermen than to any other class of men.'"[27]

Lumbermen began to alter their traditional approaches to forest management as the realities of the twentieth-century world of lumber began to come into focus. The new interest in forestry that Pinchot detected may have been sincere, but it was also somewhat naive. Lumber company executives could see as well as any other observers that the once-plentiful forestlands of the United States were being logged at an increasing rate, while the available acreage of uncut timber was dwindling. Indeed, taken to its logical conclusion, this pattern would eventually cause the extinction of their own industry. That first generation of foresters in the United States represented professional forestry as a field that provided incredible control over the natural world. Forestry was an applied science, aimed at facilitating the use of forests. Foresters carried out observations and experiments that were motivated by the specific problems and needs faced by those who were using and growing trees. Unlike fundamentally descriptive sciences like botany or geology, forestry was prescriptive. Aware that their evolving industry would soon face new challenges, titans of the lumber industry welcomed Pinchot's assurances that scientific forestry could be the key to keeping them in business.[28]

Use and Ownership: The End of the Pinchot Forest Service

The Forest Service, and the Bureau of Forestry before it, had simultaneously handled two different sets of problems. Pinchot, Zon, and their fellow foresters were often concerned with the question of forestland ownership but worked to control this through the means they had available to them: the management of forest use. The federal forester was concerned not just with scientific problems but also with the far different political and economic questions of land ownership. Foresters and politicians made decisions in reaction to pressure from lumber companies and local communities; thus the realities of the Pacific Northwest lumber industry become an important part of understanding the history of forestry both regionally and nationally. The unique ecological issues of the Douglas fir were pressing concerns for forest researchers. Planning and managing logging in the Douglas fir were pressing concerns for the Forest Service and other federal land use agencies. Furthermore, as the lumber companies reached deeper into the forest, their activities created deep concern for those who feared the destruction of wild forests would go too far. Writers such as John Muir worried that an overreaching industrial capitalism had moved into wildlands once seen as beyond the reach of human control.[29]

Some critics argued that conservation-oriented management professions should remain purely scientific and keep their distance from the political concerns of landownership. Thomas Chrowder Chamberlin, an eminent geologist and soil scientist, became concerned that conservation was becoming too enmeshed in politics to maintain its scientific focus. In an editorial in the *Journal of Geology*, he protested this "sinister current," which he feared would divert conservation from what he considered its proper function. He protested recent moves by federal land management toward controlling ownership, stating, "The conservation of natural resources centers in the scientific and technical; the right of ownership and the most desirable form of ownership center in the political and sociological."[30] He pointed out that the richest and most powerful landowners often benefited the most from scientific management, saying that "all questions of the possession and distribution of values be marshaled under extreme individuality, extreme monopoly, or some combination of individuals and corporations lying between these extremes, all are alike political and sociological in nature and . . . they leave the scientific and technical problems of conservation of natural resources to be solved on their own bases." To allow those working toward the conservation of resources to become active in regulating the patterns of ownership would be to doom these fields while they were still in their infancy. Chamberlin's editorial received wide attention in conservation circles and the popular press, just as Pinchot's view of conservation was losing popularity. The *New York Times* quoted lengthily from his piece, concluding that Pinchot suffered from this "confusion of mind," and further editorialized that "he started from a conservational premise. But he has reached political conclusions with amazing rapidity."[31]

Chamberlin's critique was protested emphatically by many foresters, who were still loyal to Pinchot as well as to Pinchot's vision of the profession as simultaneously politically oriented and research driven. They argued that the separation of ownership from use, while it was logically sound, did not make sense for the real-world management problems with which foresters were concerned. *American Forests* proclaimed that "the movement itself is fundamentally a demand for the honest and efficient stewardship of the people's property in the resources of the nation. Of necessity, it is a political movement, because it has been set in motion in response to an ethical awakening which, in turn, was brought about by a broader and more farseeing vision of economic and sociological requirements."[32] Citing Theodore Roosevelt's fusing of the natural resources with the national to create the national forest system, foresters argued that only through nationalizing forest management would it be possible for them to develop or implement any of the scientific work of forestry.

2 The Dynamics of Science, Rooted in Place
Field Stations and Forest Ecology

Thrust out into the Pacific, the shining, snow-capped peaks of the Olympic Mountains form one of the most stunning of the Northwest's many breathtaking landscapes. The Olympic Peninsula, bordered by Puget Sound and the Pacific Ocean, is home to ruggedly beautiful mountains and rain-swept forests. Grover Cleveland set aside more than two million acres of the peninsula as a National Forest Reserve in 1897, but four years later, under pressure from the lumber industry, William McKinley withdrew almost a third of that area from protection. While the high peaks attract the eye, the thickly forested hillsides below hold many charms of their own, including an isolated, distinct subspecies of American elk (*Cervus canadensis roosevelti*). The elk that roamed the low coastal forests of the Olympic Peninsula had lived apart from the other populations of American elk for so many generations that they had become a separate subspecies. Trapped on their lushly forested peninsula, however, they had nowhere to retreat as lumber companies cut ever deeper into their habitat. Their numbers dwindling as the region's population and economy grew, to many observers they seemed doomed to extinction. A 1904 bill for the creation of Elk National Park met with failure amid congressional worries about the curtailment of logging. Despite the defeat, Representative Humphrey and members of Washington's congressional delegation continued to advocate for the protection of the elk. The *roosevelti* subspecies in question had been formally described by C. Hart Merriam in 1897, who had named it in honor of his friend Theodore Roosevelt. As the so-called Roosevelt elk slid toward extinction, Humphrey appealed to Roosevelt's vanity as well as his love of hunting to sway him toward action.[1] Public demand for more careful management grew as excessive sport hunting and winter starvation decimated the herds.

In 1909 the elk's supporters finally succeeded in acquiring some level of protection in the last days of Theodore Roosevelt's time in office. As one of his final acts as president, Roosevelt designated 615,560 acres as Mount Olympus National Monument, setting a precedent for presidential use of the Antiquities Act in this manner. It must be noted that Roosevelt's impetus was the creation of a protective haven for the elk, not the curtailment of economic activity. Indeed, the parameters of the monument's designation reflected that. There was no prohibition against economic activity within the boundaries, including logging and mining. The monument comprised the most rugged and inaccessible heights of the Olympic Mountains, while the peninsula's shoreline and most of the vast forested flanks of the range remained within the Olympic National Forest. Although the preservation of the elk was Roosevelt's main intent, the boundaries of the monument unfortunately excluded most of the coastal forests that were the main habitat of the elk. This oversight, and the pressure from both activists and lumber companies, meant the debate over the Olympic Mountains was far from over. The Seattle Mountaineers began a campaign to transfer some of the Forest Service–controlled national monument to Park Service control by creating a national park. Not only would this transfer it from the Department of Agriculture to the Department of the Interior, it would also transform its management objectives from primarily industrial to primarily recreational. The Mountaineers, founded in 1907, had been modeled on the Sierra Club and shared that group's agenda of preservationist activism coupled with high-country recreation. The Mountaineers had been proponents of the original Elk National Park and continued to push for increased protection of both the elk's habitat and the high mountains.[2] The proposed park would be carved from the most rugged and remote parts of the monument. However, the drive for park status for the choicest high country was countered by industry appeals to open the rest of the monument for mining and logging. The United States' entry into World War I changed the terms of the debate by giving new urgency to industry claims. Woodrow Wilson authorized cutting the size of the original monument almost in half in the name of access to its resources, specifically its spruce and manganese ore. In a management plan issued in the spring of 1916, the Forest Service advocated creating large sustained-yield units in the forests of the Olympic Peninsula. Underscoring the agency's commitment to providing timber to the forest industry, the Forest Service appealed for the peninsula to remain under its control.[3]

By the early 1920s the lumber activity on both the public and private lands of the Olympic Peninsula was significant. However, its relative proximity to the growing urban areas of Tacoma and Seattle led to increased hiking, hunting, and other leisure activity on these lands. Wishing to protect and preserve the lands from further industrial use, many advocated not just transforming the existing monument into a national park but also expanding the borders of that protected area into the surrounding national forest. The national park plan did not meet with unanimous support. The ramifications of a curtailment of logging and lumber production within the park boundaries were an obvious concern to many. However, this was not the only concern. The elk, which were among the most celebrated aspects of the region, were still prized by trophy hunters. Of special concern to hunters was the National Park Service's policy of protection for predators.[4] Even the Antiquities Act allowed for seemingly indisputable boundaries to be redrawn for political or economic motives.[5] The moving boundaries and changing classifications of protection for the Olympic Peninsula underscored the impermanence of many land-use designations. No matter the agency, federal land-use policies inevitably changed, often in response to pressure from outside. When demand for the resources within a protected area grew strong enough, the protection would disintegrate. And once that formerly protected forest was cut, whatever pristine nature had existed there was lost.

Early in the twentieth century, then, the elk of the Olympic Peninsula were caught between the high mountain land humans wanted to preserve and the coastal forests they needed for their species' ecological function. Their predicament can stand as an example of the difficulty of reconciling the demands of land management with the complexities of a newly emergent ecological view of that land and the species therein. The complex ecological needs of species within a landscape often run counter to the demarcations and regulations of land management. But in these decades American forestry was, to a significant extent, a profession focused on managing and administering landscapes for human needs. The growing ecological consciousness of scientific foresters often ran counter to the economic and managerial demands of their profession. This chapter explores the struggle and communication between ecologists and foresters through decades when both professions became increasingly important in the United States. Some important figures at the highest levels of federal forestry came to privilege ecological knowledge, seeing the value of the new approach to the natural world. These ecological foresters questioned forestry's managing for "money trees" at the expense of whole forests. Integrating an ecological outlook into the agenda of federal forestry would prove difficult to achieve.

The Journeys of Raphael Zon

Gifford Pinchot and his friend and employer Theodore Roosevelt are the standard central figures in histories of the formation of the US Forest Service. This chapter takes a slightly different route by focusing on Raphael Zon, a close collaborator of Pinchot's during these early years of American forestry. No person better encapsulates the interactions between professional forestry and the scientific discipline of ecology than Zon, the first head of research of the US Forest Service. Zon's unique and intriguing personal and political history, and his early role in pushing the research agenda of the Forest Service and steering the field toward increasing scientific rigor, made him a significant force in determining the direction of American forest management during the critical early stages of the twentieth century. His guiding hand led the Forest Service to embrace a place-based ecological view of the forest that broke with traditional forest management. An intriguing character, he was a scientist who maintained high standards of professionalism and scientific rigor, while also being unusually outspoken about his political beliefs and personal alliances.[6]

Zon's early life led him to a unique understanding of biology, and his story begins long before his immigration to the United States in the late 1890s. He was born in 1874 in the town of Simbirsk, Russia, on the shores of the Volga River, during the reign of Tsar Alexander II. Born to liberal, nonreligious Jewish parents, he came of age in a time and place where the fallout of the emancipation of the serfs had inspired widespread leftist intellectual politicization. Indeed, in retrospect the town seems to have been somewhat of an incubator of political radicalism. Revolutionaries Vladimir Ulyanov, who was later to achieve renown

Raphael Zon photographed soon after his emigration to New York City to escape persecution in Russia. Before entering the newly established New York State College of Forestry, Zon would support himself in New York as a journalist and drugstore employee (c. 1900). Courtesy Minnesota Historical Society.

as Vladimir Lenin, and Alexander Kerensky were born in Simbirsk within several years of Zon, and all three attended the same secondary school there.[7] Zon left Simbirsk to attend the Imperial University of Kazan, a prominent and high-caliber, if provincial, university about a hundred miles from that town. The university was known for its strong tradition of scientific innovation, and members of its faculty were heralded at the time for their pioneering work in the field of organic chemistry.[8] During his student years Zon had also been politically active within a group advocating for representative government and trade unionization. Many of these activists, including Zon, were harassed by the tsarist government and arrested for their political outspokenness.

Zon's scientific training and early career laid the groundwork for his future endeavors. He studied natural sciences at Kazan and earned a degree in comparative embryology.[9] Comparative embryology, the study of the development of species from single cell to viable fetus, had become a vibrant and important field of study in the late nineteenth century. Ernst Haeckel's theory of recapitulation linked the patterns in the morphological evolution of a given species to the embryonic development of individuals of that species. Haeckel's theory thus elevated comparative embryology to a cornerstone for understanding the mechanisms of evolution, and hence a central pursuit for a generation of evolutionary biologists. Asexual invertebrate propagation was popular for embryological study, as focusing on such simple animals allowed researchers to easily isolate mechanisms of reproduction.[10] Zon's research, under the title "Sexless Propagation of Worms," was published in the *Scientific Transactions of Kazan*. After obtaining his degree, Zon was appointed to work at the famed Stazione Zoologica in Naples, a plum assignment for a young scholar. The Naples field station was founded in 1872 by a prominent evolutionary biologist and collaborator of Haeckel's named Anton Dohrn, who designed it as an interdisciplinary, international site for studying marine biology. Dohrn's ideal was to create a research environment separate from the conservative, tradition-bound zoology departments of universities. For Haeckel and others, studies of comparative embryology focused on marine invertebrates due to their short life cycle and the general ease of working with such creatures. As a result, many of the advances in understanding within embryology, and thus evolutionary biology as a whole, happened under the auspices of marine biology. By the time Zon visited in the 1890s, Naples had grown into one of the world's centers of innovation in comparative embryology. The Stazione Zoologica had also developed a uniquely creative and collaborative atmosphere as a productive site of scientific inquiry, drawing high-powered scientists from around the world. The Stazione was one of the earliest centers for the sort of rigorous, interdisciplinary, field-oriented biological study that would be central to the development of ecological thought.[11]

Unfortunately, Zon's time at Naples was cut short, apparently because he came under scrutiny back in Russia for his earlier arrests. He was called back from Naples to his homeland to answer charges of illegal trade union organizing, which was being repressed by the tsarist government. Along with many other liberal-minded Kazan students, Zon soon found himself facing a sentence of eleven years in a Siberian prison. Let out on bail, the twenty-year-old Zon fled Russia rather than face that fate. Zon and his future wife fled to Tilsit, one of the easternmost towns of East Prussia, on the militarized border between the Russian Empire and the German Empire.[12] The two soon left Tilsit to move to Belgium, where Zon again took up his studies in the biological sciences. At the University of Liège, Zon was again among scholars with a strong interest in comparative embryology and other newly important fields of biology. Presided over by zoologist Edouard van Beneden, Liège had a marine station where the work on comparative embryology was similar to that of Naples. Van Beneden became known for leading a strong research program in evolutionary morphology as well as for his multiple discoveries in cytology.[13] While studying at Liège, Zon also continued to refine and develop his leftist ideology by attending lectures in political economy and sociology at the Université Nouvelle in Brussels.[14] Zon's time in Brussels was active but brief, as the couple soon moved on to London.

In England, Zon would find a way to fuse his scientific and political interests. He was in London long enough to make extensive connections within the city's leftist intellectual elite. The couple stayed in London for a number of months, and although he did not pursue scholarship or laboratory work while he was there, the multilingual Zon worked mainly as a translator at the British Museum.[15] English socialism differed greatly from the revolutionary fervor of Russia. Free of the oppression of a tsarist regime, socialists here focused on organizing trade unions, critiquing government policy, and advocating gradual, nonviolent change.[16] Zon assisted famous labor activist Tom Mann in his organizing efforts, though his Russian experience in this field may not have crossed cultures very well. Notably, in London Zon also became active in the early, strongly socialist era of the Fabian Society, a leftist English political group. The discourse of Fabian socialism was an intellectual political critique, based in academic political thought and studies of economics. While the Fabian Society was rooted in a critique of modern industrial society, it was not revolutionary in intention. George Bernard Shaw, an early member of the society, described a "preference for practical suggestions and criticisms"[17] and an atmosphere of intensely intellectual discourse; in its critiques of market capitalism Shaw characterized the society as "the recognized bullies and swashbucklers of advanced economics."[18] Its members advocated for social responsibility and equality in the present, and a gradual shift of politics toward a socialist state. As an early historian of the society explained, "what the Fabian Society did was to point out that Socialism did not necessarily mean the control of all industry by a centralised State; that to introduce Socialism did not necessarily require a revolution because much of it could be brought about piecemeal by the votes of the local electors."[19] The Fabian Society had a number of prominent scientists within its circle, including the ecologist Arthur G. Tansley.[20]

In 1898 Zon and his wife immigrated to the United States, where he would come to devote his life to promoting federal research in the government of his adopted country. Upon the couple's arrival in New York City, the immigration authorities almost turned them away on the grounds that the young intellectual could be categorized as neither a laborer nor a tradesman.[21] Brief employment as a drugstore clerk got him through the immigration process, while on the side he wrote articles in a Russian-language newspaper for the city's burgeoning immigrant population.[22] Soon, though, Zon returned to his calling as a scientist, matriculating as a student in the inaugural class of Cornell's New York College of Forestry, where forest biology was a primary part of the general curriculum. Given that his previous scientific training and interests had been in the pure, research-driven field of evolutionary biology, this was a definite change in his career path. The American scientific landscape was different from that in Europe, and Zon had few connections in his new homeland. Finding a place in a graduate school in laboratory biology would have been difficult with a Russian undergraduate degree, no preexisting links to American universities, and very little money to his name. However, Zon's abandonment of evolutionary biology was more than merely a simple case of practical necessity. Zon had also undergone a realization during his time in England, stimulated by the Fabians' belief that science should be done in service to human needs rather than for its own sake. In time, Zon would come to describe this new belief as a conviction that forestry was a political as well as a scientific field, musing that "forestry, it seems to me, is after all only a small part of a much bigger liberal movement."[23] While the ramifications of evolutionary biology could be, and often were, used within political discourse, evolutionary biology's direct value to society was hard to fathom. Studying forest biology seemed, in 1900, to hold the promise of fusing cutting-edge biological science to specific and important applications to policy and economics.

In his journey from Russia to the United States, the young Zon was exposed to the zeitgeist of European biological science. Attracted to innovative scientific ideas and philosophies, Zon absorbed an approach to thinking about the natural world that would later set him apart from many of his less cosmopolitan American colleagues. Thus, in transitioning from evolutionary biology to forestry, Zon managed to keep his career path within scientific research but attempted to meld it in a more complete way with his strong political stance. By the time he arrived at the School of Forestry, Zon had been a part of innovative laboratories and research programs in several European countries. While his own training and experience had not been on the ecological side of biology, much less in the field of forestry, his interests and background drove him toward biological research within the field of forestry. His experiences moving through the labs, field stations, and classrooms of Europe had impressed upon him a unique sense of the value of experiment in biological science.

Zon's cosmopolitan background and rigorous biological training differed greatly from the academic background of other members of the class of 1901. The Prussian Bernhard Fernow, who had been responsible for establishing the federal Division of Forestry in 1886, presided over the Cornell school. In 1898, Fernow had relinquished his position as chief of the Division of Forestry to Gifford Pinchot to become the first dean of the School of Forestry at Cornell. As

Gifford Pinchot, then Chief of the Division of Forestry, at his desk in Washington, DC (c. 1901). Library of Congress, LC-DIG-PPMSCA-19459.

dean, Fernow arranged a course of study around the management and harvest of timber with an emphasis on forest economics. Fernow's curriculum relied on an almost wholesale transplantation of German forestry textbooks and teaching methods into the far different American context.[24] Forest biology, known within the discipline as silvics, was taught as a static field, focusing on morphological understanding of the structures and development of specific tree species isolated from their surroundings. Silvics was traditionally based in the classic approaches of old-fashioned natural history and botanical morphology. Many of Fernow's contemporaries criticized him for his apparent tone deafness to the American forests and lumber industry, and his preference for applying European management techniques suited for a much different social and cultural context. Fernow's decision to manage the School of Forestry's demonstration forest, a 30,000-acre tract in the Adirondacks, by clear-cutting the native hardwoods and replanting with nursery-grown conifers led to allegations of mismanagement and general incompetence as dean. Gifford Pinchot had been a harsh critic of Fernow both in his federal position and as an educator. Voicing his disappointment with the quality of forestry education being offered at Cornell, Pinchot used family wealth to found the Yale School of Forestry in 1900. The Yale program conferred a graduate professional degree to students who had already finished their undergraduate degrees, rather than the undergraduate degree in forestry offered by Cornell.[25]

Newly minted Cornell graduate Raphael Zon found employment as a federal forester in 1901. The young Russian brought an especially unique perspective, despite having come out of Fernow's Cornell program, thanks to his diverse training and political sensibilities.[26] From the beginning of his career as a forester, Zon was a champion of the role of scientific research for advancing the field of forestry. He had already been a believer in the importance of experimentation in biological research long before he became a forester. He understood the importance of research in forestland management from exposure to the European institutions of biological research rather than the European institutions of forestry. Basic scientific research in forest and tree biology could have its own autonomy rather than be carried out under the auspices of more practically oriented management.

In 1905, when the Bureau of Forestry was reorganized into the US Forest Service, Pinchot and Zon built a strong research program into the new administrative structure. The talented and outspoken Zon quickly became a standout employee of the Forest Service, highly valued by Pinchot, and he used this position to press for a prominent role for scientific research within the reorganized agency. Zon used his friendly relationship with Pinchot to advocate for a separate department within the bureaucracy of federal forestry, to be called the Section of Silvics. Zon conceived this as an autonomous entity under the umbrella of the bureau, where scientists could pursue topics of forest research without being beholden to the yardsticks of practical management concerns. Pinchot understood the value of forest research and was receptive to the idea that rigorous research needed more priority within an agency that on the whole had been oriented mostly toward partitioning and distributing publicly owned land. Pinchot gave research such priority in 1906 as part of his overall transformation of the old Bureau of Forestry, reengineered into the US Forest Service. The new entity included most of the elements of the old bureau but included a new, and largely independent, Section of Silvics. Further, not only did Pinchot enact Zon's suggestion but also charged him with the task of breathing life into the newly created bureau by putting him in control of it. Zon created a modern scientific base within the existing structure of the Forest Service. The Section of Silvics would need to reorganize and take stock of the hodgepodge of scientific inquiry already under way in various corners of the Forest Service, and create a framework and agenda for a more rigorous and thorough research agenda for the future. Once that reorganization was complete, high on Zon's list of desired changes was the establishment of a series of experiment stations. He lobbied Pinchot to proceed with a plan that conceived of a new element in American forest research, and indeed, federal research in general. Zon's plan took a great deal of inspiration from the biological field stations with which he was familiar.[27]

Zon became a prominent figure not only within the world of the Forest Service but within the profession of forestry in general. Unlike an academic professor, he wielded most of his power behind the scenes, and the true extent of his reach is not reflected in his research and publication record alone. In his position as de facto chief of research within the profession of public forestry, he became one of the country's most important arbiters of research priorities in biological forest research. He soon gained a reputation as one of the most knowledgeable and critical scientists in the profession. In many ways he continued to reflect the values he had absorbed during the years before he came to the United States. He became known as an intellectually rigorous, if often argumentative, gadfly within the discipline. He was not afraid to take a critical view of standard practices, even when doing so made his fellow foresters look bad. In numerous journal articles, at conferences, and elsewhere, Zon emphasized the need not just for professional training for foresters but for training that involved learning the value of scientific understanding and the methodologies of field research and experimentation. Also, he expected a much more stringent level of scientific rigor from foresters than there had been in the past. Zon believed the standards of forestry should be raised to be on par with the standards in the purer scientific disciplines.[28]

Forestry and Field Experimentation

Ecology, too, was struggling to define itself. And ecologists relied, sometimes quite heavily, on forestry for examples and explanations as they struggled to define their new science. During these years the relation of science to conservation became a topic of deep importance as both ecologists and foresters struggled to define their purpose. A segment of professional foresters found a deep intellectual resonance with this mechanistic ecology. Those foresters, drawn by the political as well as scientific message, began to alter their own forestry research in a more ecological direction. This period marked the beginning of a tension within the ranks of foresters over both the practice and the purpose of forestry research, a tension that would grow into a rift in the decades to come. Scientific discussions over the meaning and scope of ecology affected the understanding of both forested landscapes and the economic activities that took place upon them. While scientifically fruitful, this turn also began a process of lengthening the distance between the scientist and the natural world. As ecology developed as a science nationally and internationally, for foresters these advances in theory and method always played out within local environments.

One of the difficulties for pioneers marking out the bounds of a new discipline is finding acceptable foundations on which to draw for their own work. At a time when the store of truly ecological studies was still relatively small, many ecologists borrowed from other, more applied scientific fields for the basis of their work. Oftentimes they had to cast the net wide to garner any fish at all. Claiming continuity between domesticated and wild species was not uncommon, and analogies were often made between well-studied domesticated species and their poorly known wild relatives. Many research-oriented foresters were drawn to ecology for its promise of encompassing both biotic and abiotic factors into its explanations. Biological investigations in forestry had previously centered on morphological study and had been secondary to the "real work" of management and administration.[29] And in turn, ecology understood forestry as a kindred discipline. Outside the United States' restrictive professional framework of forestry, many scientific ecologists slid back and forth between forest topics and other ecological topics. Ecology in the 1910s and 1920s was often interested in finding applications, and ecologists and foresters alike often deemed forestry the best example of so-called applied ecology. The field of ecology, newly formed and in its first generations of practitioners, was enmeshed with applications for land management. This was especially true of Tansley and his mechanistic ecology that was free of some of the difficulties of the mainstream Clementsian holism. Not surprisingly, then, Tansley's articulations of ecological interrelations would get far more attention from foresters than did the dubious superorganismal explanations of the mainstream.

Forestry was at its root a prescriptive science, intended to help managers reach their goals of lumber production and forest health. Ecology was at its root a descriptive science, intended to provide satisfying answers about natural phenomena. Although forestry encompassed a lot more than just ecological study, there were many points of contiguity. As the field of ecology came into focus, some leading researchers in scientific forestry found the rigor, theory, and above all the integration of the new discipline deeply appealing. These foresters, many of whom had been frustrated by their own discipline's patchwork of goals and incompatible agendas, hoped to forge a closer bond between ecology and forestry. Ecologists were contributing research on forests, and there was some overlap between the disciplines. However, ecologists were developing a new way to view the natural world, and the applications of their work were not their primary concern. In addition, a certain class of foresters were excited by the possibilities of this new ecological thinking within their own field of study, yet the extent to which forestry would—or could—adopt the intellectual framework of ecology was uncertain. As forestry was a science whose end goals were prescriptive, the extent to which a new descriptive framework would be of use was uncertain. As their respective disciplines each grew stronger, foresters and ecologists worked to define their fields in relation to one another. Questions of the utility of ecology, even its validity as a separately defined field, often came into play. Ecologists self-consciously pointed out the value of their field to management, even as they forcefully defended their autonomy. As ecology properly included not just plant and animal species but climate and soils as well, its practitioners claimed multiple roots for their discipline, including forestry. Some foresters found in ecology a new way of thinking about forests, a way that made sense of their inherent complexity and interconnectedness. While ecologists took pains to separate their discipline from conservation and preservation, ecology still often had much to offer forestry. Ecology appealed to that subset of foresters who viewed the forest biologically, on its own terms, and believed such study would lead to true insight rather than mere profit. They adopted the analytical framework of ecology even as they applied their findings to the concerns of forestry.

In 1917, Raphael Zon published an article in the *American Naturalist* titled "Darwinism in Forestry," which showed the development of his understanding of forest studies within the wider realm of ecology. He opened with an account of the 1860 correspondence between Charles Darwin and a forester named Patrick Matthew following Darwin's first publication of *The Origin of Species*. Upon reading *Origin*, Matthew had recognized the similarity between Darwin's theory of evolution by natural selection and the theory he himself had put forward in an appendix to his own 1831 book on silviculture. Matthew had noted that the strongest and most adept individuals had the most likelihood of surviving long enough to reproduce, an effect that became magnified over generations. While Matthew's book dealt mostly with timber and trees, the well-understood relationship of timber supplies to naval power led him to make observations on colonialism and politics, and his appendix considered the ramifications of natural selection for human, animal, and plant communities alike.[30] Darwin had acknowledged that Matthew "briefly but completely anticipates the theory of Natural Selection" and remarked to Charles Lyell that "it is certainly, I think, a complete but not developed anticipation! One may be excused in not having discovered the fact in a work on Naval Timber."[31] Darwin freely gave credit to Matthew in the "Historical Sketch" that accompanied subsequent editions of *Origin*, emphasizing that Matthew's formulation of the theory appeared "very briefly in scattered passages in an Appendix to a work on a different subject, so that it remained unnoticed."[32]

Zon takes pains to explain to his readers that he is not attempting to detract from the importance or originality of Darwin's theory. Instead, his purpose is to explain why a forester would be apt to observe the struggle for existence, and through such close study of interactions in the forest formulate a theory for the basis for natural selection. The struggle was more apparent in the forest than in any other assemblage of species because of the evident competition between individual trees for the space, light, and water they needed for survival. Indeed, foresters studied the process in depth and developed sophisticated quantitative approaches to examining and describing the multiple interacting forces of competition. Because of the relatively small numbers and long life span of individual trees, the results of competition could be easy to detect. The immobility of trees over their entire life span meant that local competition was often quite intense, more so than in animal species. Further, most tree species' long life spans and exceedingly slow rates of migration meant that generations of struggle compounded within in a single site. These peculiar characteristics meant that fewer phenomena muddied the results of natural selection than in most other species. The result, Zon wrote, was that the struggle for existence could be particularly clear for tree species. The struggle was evident between individual trees of a single species, the motive force of Darwinian evolution, but also "the forest, as a whole, battles for its existence against the adjoining meadow, swamp or shrub vegetation; the old trees against the young growth that comes up under them; groups of trees of different species or of different ages against each other."[33]

Zon aimed to legitimize the scientific study of forests as a form of ecological inquiry. His intention was to show not just that forests, and foresters, figured into the history of evolutionary biology, but that studying forests could be a form of ecological science. The choice of the *American Naturalist* as the journal for this publication, unusual for Zon or for any forester, is important. This journal has long been one of the premier journals for articles on the conceptual side of ecology and evolution, publishing articles by and for those scientists most interested in the possibilities of the field. Zon prepared his article for this audience because he wished to assert a place for twentieth-century forestry within the discipline of ecology. *Origin* was universally heralded as the cornerstone of ecological thought and a continual wellspring of the discipline. By calling attention to the forester in the origin of *Origin*, Zon asserted that forestry had always been a fundamental part of ecology. While he did not articulate it, his article also served to remind contemporary ecologists of the perils of making Darwin's mistake by neglecting the rich store of research into the dynamics of forests generated by the ranks of professional foresters. It is important to note that Zon was not claiming— or advocating—that foresters emulate the investigatory style of Charles Darwin, but that their discipline consisted in understanding the exact nature of the struggle that Darwin had so persuasively defined. In fact, *Origin*'s voluminous hodgepodge of observations from natural history and husbandry was precisely the type of research Zon argued against. Instead, in this article Zon reiterated the point that the forester's correct object of study was the struggle for existence, rather than the characteristics of trees. By focusing not on the species but on the

Aerial photograph of Weyerhaeuser Timber Company logging operations in Cowlitz County, Washington (1920s). Foresters hoped blocks of uncut Douglas fir might spur natural reforestation of logged blocks. However, overhead cable yarding created star-shaped pattern of grooves on the ground of each logged patch, damaging the soils and duff necessary for regeneration. Photograph 409246; Records of the Forest Service, RG 95; National Archives at College Park, MD.

struggle, foresters extended Darwin's own work in new ways. In his conclusion Zon stressed, in italics, that *"forestry as an art is nothing else but the controlling and regulating of the struggle for existence for the practical ends of man; forestry as a science is nothing else but the study of the laws which govern the struggle for existence."*[34] Thus for Zon, all of forestry was Darwinian. The practical, problem-solving side of the discipline directed natural forces toward results that fit human needs. And the scientific side of forestry was, precisely and entirely, the study of forest ecology.

Defining Ecology and Determining Its Utility

The applications of ecology for management were many, and some ecologists were eager to highlight them. "Ecology is destined to a great future," Julian Huxley wrote in 1927. The governments of advanced nations "are waking up to the fact that the future of plant and animal industry . . . depends upon a proper application of scientific knowledge." As demands on limited resources grew, the webs of global capital laced together far-flung markets and suppliers. When an industry begins operating in a newly acquired territory, the ecological realities of that new endeavor pose new challenges. But cashing in on these new species of capital often means solving new, unknown ecological problems. Government bureaus might focus on developing specific management techniques to solve these specific problems, but Huxley pointed out that such ad hoc research ignores the larger picture. He asked whether "it is not being forgotten that behind all the detail there is to be sought a body of general principle, and that all these branches of study are in reality all no more and no less than Applied Ecology."[35] Background in the field of ecology could offer researchers unification among the specific, highly problem-oriented lines of research they were developing in relation to the demands of industry.

Ecologists also displayed a willingness to cross the boundary between natural and cultured landscapes. In addition, they tended toward mechanistic descriptions of the overall energy inherent in an ecosystem but made no pretense of describing any facets of natural history. No specific species was the focus, or even any specific place, but rather the interactions between places. The plant ecologist Edgar Nelson Transeau, for example, wrote a widely cited paper about energy exchange and the efficiency of photosynthesis using an idealized field of corn as his subject. Transeau's focus was utilitarian, based on his specific aims and the ease of calculation. Such a choice allowed the study to take place without complicating the equations with environmental factors to a level where no conclusions would be possible. An idealized agricultural field afforded the elimination of unwanted variables that would be impossible to measure and secondary to the purpose of the study.[36] Many ecologists easily commingled the wild and the tame in their ecological inquiries, but their attitude toward the wildness of their study sites was not simple to define. On the one hand, some ecological studies were concerned specifically with the description and classification of natural species and sites. Frederic Clements, for example, focused on the specific characteristics of species assemblages in his exhaustive work on vegetational climax communities. His goals were naming and describing the different sorts of plant communities, explicating relationships between communities, and comparing those communities with one another. This research agenda did not concern itself with humans or with culturally impacted sites. While the ephemeral presence of the scientists themselves was a given, and their activities and experimental interventions were allowed, no longer-term human presence was acknowledged. If these sites did contain noticeable human impact, it was minimized or ignored for the purposes of the study. Early ecologists did not necessarily delude themselves into a false sense of wildness where it did not exist, so much as take what they could from whatever tangentially related research they could find.

Ecologists did not see their role as that of conservationists or managers; they did not issue dire warnings or make prescriptions. Much of the conservation that today is seen as part of the ecologist's sphere was then considered the realm of hobbyists, zookeepers, or activists. The destructive impact of humans on the natural world could often be seen most distinctly through population declines of vulnerable and valuable animal species, and both scientists and the lay public closely observed extinctions and near extinctions of animals in the late nineteenth and early twentieth centuries. However, there was often a disconnect between the desire to save a species and the desire to save the ecological surroundings of that species. The botched attempt to stay the extinction of the passenger pigeon can be seen as an example of this. University of Chicago biologist Charles O. Whitman attempted to revive the species from a few pairs he kept in captivity, the first members of what he hoped would become the university's own "biological farm."[37] However, without knowledge of the bird's habits or ecological needs, and without true zeal for the cause of preservation, Whitman's effort failed and the species went extinct. Over those same years, reversing the near extinction of the American bison became a cause célèbre. But as Andrew Isenberg has pointed out, the bison's revival in the 1910s was a zoological reversal only. The animal has not roamed truly free for more than a century. Restoring the ecological conditions the bison required to survive — massive swaths of unobstructed grassland — was never a part of the early-twentieth-century effort to revive the species.[38] The conversion of the bison's range to grazing lands was widely accepted as an inevitable by-product of the conversion of the American West through the market revolution. Saving the bison was approached as the preservation not of a dynamic living species but of a historic artifact of a bygone era, a tourist attraction, a zoo animal. Indeed, the effort to preserve the bison was spearheaded by William T. Hornaday, a former taxidermist who was now in charge of the New York Zoo.

Those working to keep the bison or passenger pigeon from extinction were not protesting the permanent loss of the environment those species had once lived in. And the reverse was true as well: to those concerned with the fate of whole landscapes, any single species, even one as symbolic as the bison, may have been beside the point. John Muir had set this tone, when in the midst of an 1897 essay on forest preservation he wrote, "I suppose we need not go mourning the buffaloes. In the nature of things they had to give place to better cattle. . . . Likewise many of nature's five hundred kinds of wild trees had to make way for orchards and cornfields."[39] The forest itself, not any individual rare or unique species, was the important thing. Efforts to revive individual endangered species were not one and the same with activism for the preservation of whole landscapes like forests, mountain ranges, and canyons. Preserving individual species in a zoo-like setting was one step further along the path of artificiality than even Muir was willing to go. As Muir's comments betray, efforts to preserve individual species were much different from efforts to protect land from development, and in some ways the two were irreconcilable. The loss of the bison may have been symbolic of the loss of its home, but the preservation of one did not go hand in hand with the preservation of the other. Contemporary views of extinction, rooted in the close linkage activists have forged between the threat of extinction and the threat of environmental destruction, are artifacts of a more modern age.

Charles Elton represented the ecologist's stance toward these issues of conservation in his groundbreaking work *Animal Ecology*. In a section devoted to the difficulties of estimating population sizes of wild animals, he included a short but heartfelt lament for those lost. "The Arctic seas swarmed with whales in the sixteenth century, but with the penetration of these regions by Dutch and English whalers the doom of the whales was sealed, and in a hundred and fifty years they had nearly all disappeared, while a similar fate is now threatening those of the southern hemisphere." He described the pattern of reduction, even extinction, of animal populations worldwide in the wake of human impact: "Almost everywhere the same tale is told—former vast numbers, now no longer existing owing to the greed of individual pirates or to the more excusable clash with the advance of agricultural settlement." However, he observed, "at the same time there is in many cases no reason why animals should be reduced in numbers or destroyed to the extent that they have been and still are. From the purely commercial point of view it often means that the capital of animal numbers is destroyed to make the fortune of a few men, and that all possible benefits for any one coming later are lost."[40]

Having said that, however, Elton refrained from advocating any action. In cases of extinction "it is not much use mourning the loss of these animals, since it was inevitable that many of them would not survive the close settlement of their countries." While the threat of human-caused extinction threatened many species, "enlightened governments are now becoming alive to this fact, and measures are being taken to protect important or valuable animals. It seems, then, that man is beginning to rectify some of his earlier errors in destroying large and interesting animals, and that the future will in certain regions show some approach to the original condition of things before man began to become over-civilised."[41] Elton clearly distinguished between the ecologist, whose job it was to study these wild species, and the conservationist, whose job was to manage resource use. In 1925, lacking a university position, Elton found employment with the fur-trading giant Hudson's Bay Company, researching the ecological reasons for fluctuations in fur-bearing animal populations. His role was to explain the reasons for those fluctuations, the basic ecology, not to make any prescriptions for how to maximize company profits or maintain stable resources. Ecologists might note declines in population sizes and theorize about their causes, but they did nothing to limit those declines.[42]

While forestry researchers tended to design projects around large, familiar, well-developed field sites, such as university forests and the Forest Service's forest experiment stations, academic ecologists of this era often did not. Historian Robert Kohler described ecologists as being uneasy conducting experiments in the field because of the complexity and specificity of the natural world in contrast to the laboratory. While this might have been true for ecological researchers employed in academic biology departments and surrounded by lab scientists, it was less relevant for foresters and other field researchers employed by forestry schools, independent research organizations, and government agencies. Such ecologists were part of a long tradition of field experimentation. Other factors seem to have also been at least as important in this. Much forest research took place on lands devoted to research and partially or fully controlled by foresters, a luxury unknown to ecologists at that time. In both Europe and North America, foresters could devote significant sections of forest to rigorous protocols of variable testing. Foresters enjoyed significant funding in comparison to their ecologist colleagues and hence could afford to think on larger scales. Since the timescales of forestry experiments could be quite long, years or even decades, foresters did not work as frequently in highly competitive academic settings. Academic ecologists could hardly expect to commit to experiments that would not be completed in the time span of a tenure track or a doctoral dissertation and hence gravitated toward research that more easily fit the rhythms of university assessment.

Foresters did not necessarily regard the natural world with the same detachment as ecologists did. With less of their intellectual effort invested in making absolute claims about untouched nature, foresters were freer to create experiments in the field than ecologists were. This generation of ecologists did not see themselves as conservation activists, although they did view their work as generally useful for various conservation problems and concerns in resource use. Indeed, the impulse to separate the new field from the tradition of natural history may have led some ecologists to adopt an even more noncommittal stance on the degradation of the natural environment. Ecology prescribed, then, but did not advocate. Furthermore, ecologists not only understood that humans were a part of the natural world but observed that the spread of human populations and industries became at times a dominant presence in that world. Notably, however, that understanding did not necessarily lead to a condemnation of humanity or industry.

"The Very Soul of Ecology": Field Experimentation and Dynamic Ecology

Pinchot's appointment of Raphael Zon as head of research came as many American scientists were debating and reassessing the methods and motives of field-based biological research. Zon's decisions to root forest research in landscapes and to prioritize ecological approaches reflected these larger changes within field biology. Zon's allegiance to a dynamic ecological approach to forest research, and Pinchot's support of that approach, were evident in the particularities of their establishment of the Section of Silvics. Specifically, this is seen in Zon's push to establish field research stations, which began almost immediately after the Section was brought into existence. Zon and Pinchot's assertion of the importance of research showed their belief that forestry could be a part of mainstream American science, rather than a profession serving the furtherance of the lumber industry's profits and goals. The focus on forest experimentation as the means to that end demonstrated that their scientific thinking was in line with the newly developing field of ecology as the most promising context for understanding forests. At the core of this conception was the placement of scientists in situ, away from the bureaucracy of their place of employment and in the midst of their subject of study. Such sequestration, in places of welcome exile where time was devoted to fieldwork, would come to define serious ecological science.

Zon's vision of the way forward for forest research would reflect the new field of ecology more than traditional forestry; hence he desired a model for forest research that looked forward in its methodologies, not backward. The most convenient historical precedents for structuring research agendas at the forest experiment stations would have been the European academic forests and American agricultural experiment stations founded in the nineteenth century, yet Zon had little interest in them. The Prussian forestry schools' experiment stations were centrally controlled and associated with the state rather than with universities. Large-scale, rigorous experimental field trials had been under way in the demonstration forests of the largest European forestry schools for generations. Indeed, such experimentation was an integral part of forest research. The situation was similar at the American agricultural stations. Data collection and experimentation in plants and soils were extensive, but not systematic. In both the European forestry schools and the American agricultural stations, experimental methodologies were governed by long-standing traditions of land stewardship rather than by the procedures of rigorous scientific inquiry. They controlled large areas of land that was used for experimental trials, usually undertaken in response to demand for answers to very specific and localized problems or management issues. These institutions had been founded to facilitate research into the study of organisms in landscapes, but they had been organized with the goal of facilitating societal use of economically desirable species. By the beginning of the twentieth century, the scientific philosophy inherent in these institutions' organization and priorities, rooted in an organism-based approach, was increasingly out of step with the ecologically minded forefront of American field biology.[43]

Zon took little inspiration from these old-fashioned precursors of government experiment stations, instead choosing to model the American forest research stations on biological field stations geared toward the interests of academic scientists. European academic forests and American agricultural experiment stations were the most obvious historical precedents for structuring research agendas at Forest Service experiment stations, and they were often mentioned in this context by politicians in the media but were only marginally important in the actual design. The research agenda in Zon's new federal forest experiment stations focused on improving basic scientific knowledge of trees and their ecological place in forest landscapes. Thus, a new, innovative form of field experiment station, reflective of the vibrant young field of ecology, would serve as a better model in both form and function. The Carnegie Desert Laboratory, as I will explain, was the first glimpse of a new breed of American experiment station devoted to research specifically in ecology.[44]

Near the end of the nineteenth century, biologists were moving generally toward a more rigorous approach to fieldwork. As scientists adopted an ecological approach to the study of organisms and their surroundings, they sought to distance themselves from the specimen collectors and enthusiastic amateurs of natural history. As ecology took form as a discipline and an interpretive approach, academics took pains to design research programs that were far more rigorous and formalized than those of the older discipline. Field and experiment stations became part of these scientists' endeavors to impose a layer of academic structure onto fieldwork, to normalize the chaos of nature. Zon's involvement with the creatively fertile atmosphere of the Naples Zoological Station augmented his understanding and appreciation of the value of institutions devoted solely to intensive scientific work. Although the Naples station had been the first of its kind, by 1906, other examples of this trend could be found, even within the United States. During the era of Zon's visit the Naples station had been known mostly as a center for morphological rather than ecological research. The scientists were seeking to understand the workings of specific species of interest in order to gain insight into larger questions in evolution, anatomy, and physiology. The proximity of the Bay of Naples had been useful primarily to ensure a constant supply of choice specimens of the marine species so useful for the study of embryology and anatomy. However, the Naples Zoological Station quickly assumed prominence for biologists who were examining nature on larger scales: whole populations of a species, communities made up of myriad species, and the nonliving forces acting on all of those species together. The Woods Hole Marine Biological Laboratory, founded in 1888 along the coast of Cape Cod, was modeled on the Naples Zoological Station. The two sites were similar in their relationship to the landscape and in the communities of scientists who attended. However, Woods Hole's administration was a joint effort of several major universities and the federal government. Also, Woods Hole did not limit itself to one biological discipline but encompassed many scientific approaches to studying the ocean. Rather than focus energies and facilities on deeply exploring a single discipline, its founders conceived of Woods Hole as a way to bring scientists of various specialties together.[45]

As Pinchot and Zon were working out the role of research in the new Forest Service, a new model for studying ecology in the field was being built. Beginning in 1902 the Carnegie Institution of Washington, as part of its broader goal of supporting American science, established a series of field research stations with the goal of developing scientific research into chosen American environment types.[46] Over the next three decades the umbrella of Carnegie science funding would come to extend over a number of mountain, desert, and coastal research sites. Some of these, including the remote Marine Laboratory founded in the Dry Tortugas south of Key West, were established specifically to encourage field biologists to research rare and little-known landscapes. Early ecologists heralded the Carnegie Desert Laboratory, on the outskirts of Tucson, as the ideal place to foster the next phase of ecology. Many thought this could jolt the field out of the elaborate yet ultimately uninteresting synthetic studies that were so common. With the budget, independence, and focus to support ecological innovation, the Carnegie Labs could help the field move toward deeper understanding of the interactions between living things and their relationship to the struggle for existence. At its founding the Desert Lab was heralded as an institution that would allow, through its focus on field studies of plant physiology, the discipline of ecology to break free of the habits that were holding it back.[47]

The Desert Laboratory had been built to fit the needs of the new field of ecology, and the lab in turn inspired ecologists with its devotion to the new. Understanding the role of the Desert Lab in the field of ecology can show how the forest experiment stations were constituted to have a similar role. In its early years the Desert Lab did more than just provide a place for research; it did much to set new agendas for the field of plant ecology as a whole. For example, in a paper in *Science* entitled "Cardinal Principles of Ecology," botanist W. F. Ganong heralded the newly created Desert Lab as a possible catalyst to help the field of ecology move into more sophisticated types of investigation. Ganong observed that ecologists had theretofore busied themselves with the cataloging and interpretation of the vegetation, animal life, and climate of study sites; in other words, plant geography. To move forward as a science, ecology would need to begin to investigate "environmental physics and adaptational physiology,"[48] the interactions between species and their surroundings. This study, especially regarding adaptation by a species to the realities of its environmental conditions, he heralded as "the very soul of ecology."[49] Moving forward required a revolution in methodology, the development of new instrumentation, and most of all the establishment of dedicated research institutions like those of the Carnegie. As he called for new methodologies and new instruments, Ganong also underscored the uniqueness of the Desert Lab's complete focus on research. It meant ecology was no longer the purview of "busy teachers who can give to them only a vacation leisure and a scanty equipment," instead providing for "trained investigators who, with ample time, expert assistance, and properly equipped field laboratories, can give themselves wholly to [research]."[50] The mode of ecology fostered by the Carnegie Institution of Washington's field stations was inspirational to those who aimed to improve the rigor and focus of American ecology.

Leaving Behind the Common Good: Changing Goals in Ecology

Founded on opposite sides of a ridge in the Pike National Forest, the Fremont Experiment Station and the Alpine Laboratory represented two contrasting modes of field study. The Alpine Lab, established by Frederic Clements in 1900, had been conceived as a residential site for in-depth summer research by him and his students.[51] The work the Clements group generated based on ecological research at the Alpine Lab and elsewhere propelled him to prominence by the end of that decade. In 1908 Clements, by then a professor of botany at the University of Minnesota, met Pinchot at a forestry banquet in Saint Paul. In conversation he learned that Pinchot and Zon had been working on a plan to develop a series of experiment stations. Clements reacted enthusiastically, seeing potential for collaboration. He invited Zon out to visit the Alpine Lab, to see what promise the Pikes Peak area held for research as well as to show off his own facilities on the mountainside. Clements made clear to Pinchot that he would welcome a federal experiment station as a neighbor to his own institution.[52] In years to follow Clements would benefit from association with both the Forest Service and the Carnegie Institution, although his Alpine Lab maintained autonomy from both Carnegie and the Forest Service sites.

The Fremont station was one of two sites to be founded in 1909 as the first field stations of the Division of Forestry, along with the Fort Valley Station near Flagstaff, Arizona.[53] During this embryonic stage of the federal forest experiment station, Clements was briefly an official collaborator with the Section of Silvics.[54] The collaboration consisted of a study of lodgepole pine in the Rocky Mountains. The study was conducted over the summers of 1907 and 1908, based out of the Alpine Lab facilities. Clements was later listed as an official collaborator with the federal program of research for this work, but that appears to have been a post hoc arrangement not representative of a formal research agreement. The study involved topics that were of great interest to federal forest management, namely the recovery of lodgepole pine forests following fire. Clements maintained a utilitarian slant in the study, with recommendations about how to manipulate the forest. "It is by means of fire properly developed into a silvicultural method that the forester will be able to extend or restrict lodgepole reproduction and lodgepole forests at will."[55] The study, published in 1910, contributed to the understanding of the lodgepole, but the work itself was not conducted with federal support or facilities. Nonetheless, Clements seemed to be interested in fostering a more formal relationship with the new Fremont station in the future. Zon even went so far as to describe Clements as "very much interested in our present studies."[56]

While Clements nominally collaborated with the federal studies during the early years of federal forest research, his style did not leave an indelible mark on practice. Indeed, Zon was sour on Clements's interpretations of ecology even while they were officially collaborators.[57] A significant exchange happened in 1908, when Zon veiled his distaste for Clements's field methods within a review of another ecologist's article on methods for studying the light requirements of forest trees.[58] Finding experimental methods that would allow direct measurement of dynamic ecological functions, such as light uptake, was a priority for the Section of Silvics. Zon declared that "we have reached a point where mere impressions or purely empiric knowledge is insufficient. We must actually measure the physical factors and know how to do this. . . . Prof. Clements' photometer is nothing but a modification of Wiesner's photometer, and he may want to say something in regard to Zederbauer's criticism of it."[59] Clements responded to the criticism, saying that he found the paper in question "unconvincing from the theoretical and the experimental standpoint, in spite of the fact that I have admitted all along that we must some day decide whether light has a qualitative effect in nature. I determined last summer that the time had come for taking up this matter . . . and have arranged to carry out several series of experiments on the absorption of leaves and the spectrum analysis of forest light." For Clements, though, examining the process of uptake was not a pressing need, given his faith in the methodologies he already had in use. "These seem largely academic to me, however, in the light of practically all of my own work, and that of several students, especially that of Mrs. Clements in the quantitative study of leaf structure."[60] Zon wrote back, protesting that he thought "the question involved is one which should be investigated before we proceed with our light measurements in the forest." Whatever failings the Zederbauer paper had, his critique of the method used by Clements was important. "The accuracy of all methods based on the measurement of the chemical rays, it seems to me, depends on whether or not the diffuse light of the forest is of normal composition. Wiesner's experiments on this question do not seem to be conclusive, and until more convincing experiments have been made, the measurements obtained by this method will naturally be open to attack."[61]

Such differences over methodology are not as minor as they might seem at first glance. The debate over instrumentation was, in fact, a debate over which entities or factors should be the central focus of ecological investigation. Many acknowledged that the most promising avenues of research in ecology were those that made use specifically of physiological modes of inquiry, and hence were able to delve into direct questions of the dynamics between plants and their environment. The eminent Henry C. Cowles, for example, by 1908 considered the use of physiological methods indispensable in the field, observing that "it is coming to be realized that the problems of physiology and ecology are essentially identical. . . . It is the exact methods of the laboratory carried into the field that give promise of the solution" of ecological questions.[62] Such inquiry required intelligent use of the instruments and methods of plant physiologists. The instruments Clements favored tended to gather information about the biological and physical entities that made up the ecological whole—attributes of the plants and the environment in which they existed. The physiological instruments Zon favored were newer, and often more expensive and less universally available.[63] However, these instruments were more valid, as they were designed specifically to gather information about the dynamic interactions between plants and the environment, the very action of the struggle by which plants survived and thrived. Through the oblique language of a disagreement over the methods of two other proxy scientists, Zon and Clements were playing out their much larger difference of opinion about the essence of ecological practice.

Zon set up the forest experiment stations under the direction of scientists who shared his views about the importance of dynamic-analytical questions in ecology. Clements's early overtures of a collaborative relationship between the Fremont station and the Alpine Lab did not amount to much. One of his earliest allegiances came with the appointment of Carlos G. Bates as the first director of Fremont. Bates was, like Clements, a graduate of the University of Nebraska program, but despite their common training Bates's methods and goals in ecology had diverged greatly from those of Clements. Bates was a believer in the importance of innovative physiological instruments for furtherance of ecological studies. In 1909, Bates initiated the first American watershed study in the form of an experiment on the relation between forest and stream flow. This study would last for sixteen years, based on a site in the Rio Grande National Forest. His research designs were complex and ambitious, placing serious demands on the Section of Silvics' budget and equipment. Bates and Zon, both known as opinionated, hotheaded, and intelligent researchers, collaborated together often. The two men shared a strong conviction that sophisticated instruments were of primary utility for investigations into forest function, which would later be reflected in a widely read 1922 Forest Service publication, the 208-page *Research Methods in the Study of Forest Environment*. Zon's installation of Bates across the ridge from Clements's lab was strategic. Having a researcher as strong as Bates in residence limited Clements's influence on the Forest Service research goals and philosophies that would develop at Fremont.[64]

Zon's stance is especially important, and unusual, given the implications of these two modes of ecology for the management of American forests. The juxtaposition of Fremont and the Alpine Lab illustrates the tensions as ecologists struggled to define the field's scope and methods. While some scientists questioned the utility of timeworn techniques and approaches, others worked to perfect those very same methods of inquiry. However, these ecologists, although they disagreed with one another, were not simply sorting themselves into two opposed sides, each with internally consistent ideals. There was much inconsistency and dispute among the people within each group, and a number of ecologists worked in both modes or fluctuated from one position to another. Raphael Zon's stance on research priorities and methods, and hence his view of ecology in general, were thus reflected in the personnel, research priorities, and methodology of the first generation of forest research. For more conservative American foresters, the Clementsian characterization of the climax community was the default mode of describing and explaining natural forest. While Nebraska grasslands were often used as the model for Clements's theories, discussion of forest climax stability was common among Clementsian ecologists. Clements's descriptions of the inevitability of the succession-and-climax process seemed borne out by the stability of mature forests in particular landscapes.[65]

Zon's decision to step away from the appealing model of the Clementsian climax was based on his conviction that the science that underpinned that model was faulty. In order to fully appreciate what was at stake in this reorientation, it is important to understand the deep appeal of the Clementsian climax for those who managed forests at the turn of the twentieth century. One of the biggest problems that faced American forests was the ultimate fate of forestlands that had been transformed by intensive logging. The Clementsian model suggested that the land, even denuded of all vegetation, would begin the normal predestined process of vegetational succession. It would heal itself from the injury, in a reliably predictable process that was determined by its climate and other site factors. Thus, the land would eventually return to its rich state without any significant managerial intervention, become covered once again by the lucrative climax forest, and eventually be ready for another round of logging. The idea was comforting for the lumber operators, who were thus not culpable for permanent destruction. It was convenient for the managers, who could thus transfer to nature's capable hands the project of rejuvenating the damaged forest.

In addition, the inevitability written into the Clements succession-and-climax model allowed forest managers to believe they could control nature.[66] They could entertain the belief that the process was predictable, and therefore controllable. Although the vegetation changed on its own, it was possible to predict what the next stage would look like, and what the climax would look like as well. If this were true, it would take much of the uncertainty out of forecasts for the long-term health of American forests. Furthermore, if a given forest did not appear to be progressing along a desired trajectory, it should be possible to find methods to fix or correct that trajectory through intervention in the stages of succession. Clementsian forestry should thus be able to ensure the establishment of a particularly lucrative climax forest type through judicious intervention in earlier stages in succession. This might take the form of modification of the physical environment, or the introduction or control of certain secondary species of importance to the pattern of succession. Philosophically, Clementsian ecology was in accordance with conventional forestry, since both allowed scientists to act in direct service to human needs.

By denying a place for either Clements or his research style within early federal forest experiment stations, Zon allied federal forest research with the dynamic-analytical flank of ecology. Zon and Pinchot constructed a role for research in the Division of Forestry that reflected their own interests in forest ecology. Zon pressed for scientific study of the forest that probed ecological questions about the dynamics of the struggle for existence. The program of forest research Zon designed was meant to be independent of the administrative and managerial needs of the larger federal bureaucracy. The researchers would be given free rein to investigate the forest with the aim of basic understanding, not practical application. While this version of ecology did not lend itself easily to the prescriptions and predictions of the Clementsian model, the reasoning behind dynamic-analytical studies seemed more scientifically promising. This would first have implications for the first years of the forest experiment stations, setting the agenda for experimental design and long-term research. Many other researchers, drawn also to the dynamic-analytical approach to ecology, supported and contributed to this research agenda. As Zon maintained positions of power over research in the 1910s, both as the director of research and as the editor of the *Journal of Forestry*, this trajectory continued.

Conclusion

The experiment station was Zon's ideal of forest research and can be viewed as a reflection of his forest agenda. Field stations proved to be ideal places for furthering research in federal forestry. In a scientific profession without the benefit of the university setting, field stations served an analogous purpose by providing facilities where differently trained and differently focused scientists could work side by side on separate but related research programs. While the location of a field station might govern the particularities of subject matter somewhat, privileging the ecosystem in which it was set, multiple approaches could be under way at once. The forest experiment station thus became pivotal as a conduit for understanding the forest in an ecological context. The early years of ecological research, when the boundaries and agendas of the discipline were still being set, demanded cross-disciplinary collaboration. Also, forest experiment stations lent themselves to understanding a particular site as a complex array of interlocking systems, including everything from climate to human use. Forest science could evolve to be increasingly concerned with the relationships between living things and their abiotic setting, inspired in part by forestry's reliance on in situ study and experiment. American federal forestry became progressively more focused on understanding and managing the interactions between forest health and the demands of economics and culture. Federal foresters' early understanding of the importance of managing the health of watersheds, for example, was a reflection of both their awareness of ecological interconnectedness and their vision of a forest resource that included human presence.[67]

Despite a healthy amount of introspection among ecologists of the time about the proper goals and uses of their field, the difficulty of maintaining an analytical program of research while simultaneously serving the needs of the public remained. The issue was one that would continue to bedevil Zon and others who strove for research agendas that were both ecologically sound and socially useful. It is important to note, at this point, that there is more than one way for science to serve societal needs, and that there are many different competing sorts of scientific needs that can be met. Indeed, many forms of scientific inquiry do not serve societal needs in any direct way. Ecology did not confer direct utility for the stated needs of forestry in the way that other modes of research might have done. Finding a way to make this science relevant to society became a struggle for foresters that will play out in the second half of this book. For foresters, the difficulty of being both rigorously scientific and socially relevant remained. The inherent tension within an ecologically oriented forestry would incubate within the discipline, causing disputes and differences for years and inspiring a full reassessment of the field. In the 1930s, the philosophical difficulties of this contradiction, intensified by arguments over industrial demands on Douglas fir forests, would create a deep rift between foresters that would permanently divide the field.

3 On the Ground

Ecological Experiments and Philosophical Refinements

The Wind River, fed by high-mountain snowmelt, runs south from the flanks of the northern Cascades to join the stately flow of the Columbia. By the turn of the twentieth century, the steep slopes and rugged valleys along the river had been inhabited by Indians, trappers, loggers, and farmers and now were held by the federal government. In 1905, the lands around the Wind River were included in the massive acreage of forestlands transferred out of the management of the General Land Office into the hands of the newly created US Forest Service. Within a few years, the Forest Service established a research area in a valley along the river, a place to begin study of the increasingly important Douglas fir forest. The Wind River site quickly evolved from a simple seedling nursery into the Northwest's foremost site for federal forest research. The Wind River Experiment Station, as it came to be known, was located near Carson, Washington, west of the Pacific Crest and south of Mount Saint Helens in what is now the Gifford Pinchot National Forest.[1] Close enough to Portland for easy travel, the station was still remote enough from urban and agricultural developments to be truly a part of the Cascade forests. The proximity to the Columbia River had facilitated early logging and exploration of the area, and the valley was well mapped and described by the time the Forest Service moved in.

This valley became a site of forest experimentation even before it was officially designated a forest experiment station. Thornton T. Munger, a newly minted Yale forester, was sent out from Washington in 1908 as Raphael Zon's sole representative of the Section of Silvics in the region. Munger had a massive, if nebulous, goal: filling in some of the many blank spots that still existed in the scientific understanding of the Douglas fir forest. One of his earliest projects was establishing "permanent growth plots" in 1910, scattered throughout the Wind River valley. Hiking through the forests along the Panther Creek and Trout Creek tributaries of the Wind River, Munger found areas where fires had killed off mature trees. The openings from the fire had resulted in the growth of a new generation of seedlings by creating favorable conditions for the shade-intolerant young Douglas fir. He marked off acre-sized squares and tagged every tree that grew within them. For each tree he recorded the size and age and committed the Forest Service to returning to monitor their growth in an open-ended study of

Wind River Forest Experiment Station, with nursery in the foreground and experiment station buildings arrayed behind. Civilian Conservation Corps men labor on the nursery seedbeds (1933). Blown-down trees litter the slope behind the buildings. Photograph 95-GP-2073-280406; Records of the Forest Service, RG 95; National Archives at College Park, MD.

natural Douglas fir forests. As Munger established these permanent sites, others worked on clearing land and building structures for a tree nursery in the valley, to produce seedlings for future reforestation work. With the establishment of the permanent ecological studies and the nursery facilities, the Wind River area was transformed into a nascent center of Forest Service research in the Douglas fir.

The Wind River Experiment Station was part of the first wave of forest research stations to be established following Zon and Pinchot's push for American forest research. The Forest Service had committed to understanding the forests on a scientific level, which in the Pacific Northwest meant working from very little preexisting scientific knowledge. The station aimed to develop scientific knowledge about the Douglas fir forests both to improve the economic profitability of the region's lumber industry and to maintain the long-term health of the forests. Researchers integrated logging and reforestation into their studies of the forest as well, developing studies of the human role in the Douglas fir forest's long-term stability and response to disturbance. In 1912, Forest Service researchers began a series of experimental plantings to determine the varying environmental needs of valuable tree species. In 1918 a long-term experiment was established to examine how quickly the forest recovered from some clear-cut logging in the forest the year before. The experiment station would be a new avenue toward compiling the sorely needed knowledge about the basic ecology of the Douglas fir forests, as well as a way to investigate possible methods of logging, reforestation, and forest management as the lumber industry became an increasingly important part of the region's economy.

Ecological thinking led foresters in new directions, and the Wind River Experiment Station became the site for some important early work integrating ecology into forestry. As chapter 2 demonstrated, Raphael Zon and Gifford Pinchot were inspired by ecological modes of inquiry as they shaped the federal plan for forest research in the early years of the US Forest Service. This chapter examines the effects of that ecological approach to forest science on federal foresters in the late 1920s. Such an ecological approach offered not just different methodological approaches to forest science but also a different way of understanding the role of scientists in forest policy decision making. This chapter describes how some foresters made use of ecological innovations to understand the forest in a deeper way than before. Munger, Leo Isaac, and others stationed at Wind River used experimental ecology to find reliable answers to the pressing questions underlying forest management issues in the Douglas fir region. Studying the forest as a whole, a site of complex interactions and long-term developments, was made more reliable by viewing the forest through an ecological lens. In the history of the beginnings of the Wind River Experiment Station we can see the agendas of Forest Service research for understanding both the basic science of the Douglas fir and for finding ways to integrate the lumber industry into western forests. Work at the station was conceptually situated at the intersection of ecological science and industry-oriented problem solving. This forest can be a touchstone for understanding the development of particular forest science agendas, in both Washington, DC, and the Cascades, in the years after it was established.

When forests were conceived as ecological entities, new directions of inquiry opened for researchers, but that ecological approach also made it difficult for those researchers to integrate their findings into the framework of federal forest management. Ecological research goals had the potential to clash with the demands of the US Forest Service. However, the potential also existed to merge ecological forest science with the sense of social utility that had traditionally characterized American public forestry. This chapter examines one such person, Bob Marshall. Marshall, famed for his later interests in forest recreation and wilderness advocacy, built them upon a foundation of ecological research. Most historians of the wilderness movement have neglected Marshall's professional identity as a scientist, and most historians of forestry have not examined the profession's role as an incubator of wilderness activism. For some foresters, scientific interests became increasingly allied with the field of ecology, focused on the interactions between trees and their environments in natural forest settings. While ecological studies were scientifically intriguing, they often lacked any goals of direct social utility. Forestry's potential to enact social as well as ecological change increasingly occupied Marshall both personally and professionally. He drew close to other foresters who similarly linked the social and the ecological, including Zon and Pinchot. Their ultimate goal was to reinvigorate the social conscience of the American forester. In their eyes, the conception of nature inherent in the modern science of ecology was closely linked to wilderness preservation. This dynamic ecological vision of the world allowed Marshall and his compatriots to shape such a different view of wilderness from what had come before.

This chapter, like the last one, moves from the Douglas fir into an exploration of how that tree impacted the development of American forestry. The experiments at Wind River were manifestations of Zon's vision of forestry as a form of dynamic ecological study of plant populations. Bob Marshall, who served as an assistant on those Wind River experiments, came to embody the forester as ecologist. As Marshall became more interested in ecological research, he and his allies began to question the profession's emphasis on aiding the lumber industry. Beginning in graduate school, Marshall defined himself as a scientist and as an ecologist, a sense of himself that would in turn influence his conception of social utility for forestry. Alienated from the mainstream of their profession, this radical coterie became a small, but vocal, new faction within the ranks of professional forestry. These foresters formulated an agenda for forestry that would, over a span of several years, shake the profession deeply and ultimately divide its ranks.

Ecological Practices: Finding Answers in the Douglas Fir

The first director of the Forest Service's new Wind River Experiment Station was a young forester named Julius V. Hofmann. Hofmann was still finishing his doctoral thesis at the University of Minnesota's School of Forestry when he was hired. One of Hofmann's most pressing duties at the new research station was to investigate possible methods of reforestation of the Pacific Northwest, a problem just coming to prominence as the new clear-cutting regimens quickly laid bare the forests. As a further complication to his assignment, artificial reforestation methods for cutover national forests were distinctly out of favor, both among the perennially cash-strapped Northwest loggers and within the Forest Service administration. The current legislative mandate for the Forest Service at the time was the 1897 Forest Management Act, which precluded the Forest Service from enforcing any artificial reforestation efforts, although enforcing logging techniques that would enable or encourage natural reforestation was still legal. Thus, Hofmann was encouraged to explore the natural regeneration of Douglas fir, while still keeping an eye toward development of cutting regimens that would facilitate quick regeneration.[2]

Within five years Hofmann had developed a theory of Douglas fir reproduction by which the full regeneration of the forest required almost no action on the part of either lumber companies or the Forest Service. This feat is even more impressive given that there was almost no previous research on the reproduction of Douglas fir to suggest possible methods of attack. Hofmann's seed-storage theory, laid out in a single lengthy 1917 paper, hinged on the character of the duff, the layers of decomposing needles making up the topmost stratum of the forest floor. Hofmann noted that within several years following forest fires or logging, seedlings often appeared hundreds of feet away from the nearest living mature tree. This was too far, he believed, to be the result of seeds blown by the wind from the seed trees into the treeless areas. He concluded that the seed had instead been dropped by trees before the disturbance had swept through the area, and had lain in the duff since that time. He stated that "the duff contains a large number of germinable seed, which might remain dormant there for a number of years and which evidently germinates and results in a dense stand of young growth as soon as the forest is cut down or burned over and light and heat are admitted to the ground."[3] Thus all forest duff in the Douglas fir forest contained large amounts of seed, from many different years, all waiting for the opportunity to germinate. Hofmann strengthened his argument by describing a burn where the fire had in places burned so uncommonly hot on the ground as to destroy the layer of duff entirely, leaving only the mineral soil that had lain beneath it. "Wherever the duff and litter were not burned out of the forest floor, there developed an area of more or less dense reproduction," he wrote. He concluded that in the locations where the intensity of the fire had burned away the duff, the seed had burned with it, while in the places the fire had burned at a more moderate intensity and the duff had remained, "seed must have been produced and stored in the forest floor before the fire and have retained its viability through the fire."[4] Hofmann saw the activities of rodents as pivotal in this storage of seed, through their habit of caching large amounts of fresh seed within the duff layer as a food source for the winter months.[5]

Hofmann's paper relied almost entirely on deduction from observation and from the synthesis of results obtained from other researchers' experiments on various plant species. This is most evident when Hofmann discusses the viability of Douglas fir seed after a year or more under the duff. Hofmann described the duff as the ideal storage medium: "It is not at all surprising that germination should be delayed under the forest cover. The cool shaded layers of leaf mold and general duff of the forest floor, which in the virgin Cascade forests seldom feel the warmth of the sun, constitute an ideal storage medium. Under conditions so unfavorable to germination and so favorable to its retardation, it can easily be imagined that the germination of forest tree seeds can be delayed to the limits of their various powers of dormancy."[6] He admitted that he had not succeeded in getting Douglas fir seed, or any other forest tree seed, to germinate after being retrieved from the forest floor. Furthermore, an attempted experiment to bury fir seed for set periods of time and then unearth and study the seed was aborted after rodents disturbed the experimental containers after the first year of the study.[7] Hofmann relied instead on the work of other researchers to determine the viability of seed. For example, he cited unpublished research from Idaho's Priest River Experiment Station, where western white pine seed recovered from the forest floor was found to still be viable. He also cited unpublished small-scale reports from Forest Service nurseries in which germination of redwood, incense cedar, sugar pine, and a number of other western forest species had been delayed by several years. Such studies did not, however, address the question of wind-borne seed entering the nursery grounds from elsewhere, a fact that Hofmann does not highlight. The most persuasive studies on the long-term viability of seed, and the only ones that had been published, did not involve forest seed at all. While Hofmann cites the work of six different researchers on long-term seed viability, he acknowledges that none of them were studying forest tree seeds.[8] As Leo Isaac would later describe dismissively, "he got that idea from some viable wheat seeds that had been taken out of a tomb somewhere in Germany several hundred years after they were stored there."[9] Ignoring the ecological context and specific attributes of Douglas fir was discordant with the sensibilities of ecology, although not an unusual sort of inference to make when working within an older, more morphologically based understanding of forestry.

While Hofmann knew that Douglas fir was not the climax species of the Pacific Northwest forest, his theory of seed storage nonetheless relied heavily on the influence of the succession-and-climax framework. The second-growth forest would return and be exactly the same as the first growth, but only if humans neglected the cutover site entirely. If left alone, Hofmann insisted, the seed stored in the duff would replicate the lush cover of fir more quickly than any artificial reforestation project could do. Indeed, even the mild intervention of a slash burn, to clear the cutover area of debris, would impede the forest's natural regeneration. There is an echo here, in such complete faith in the seed's ability, of the more providential ideas of the succession-and-climax model, by which any disturbance is only a temporary setback on the march to the climax. However, Hofmann also believed that once understood, the succession-and-climax pattern in vegetation could easily be manipulated. Hence, while he believed that the reestablishment of the Douglas fir forest would be best accomplished by letting nature take its course, he also believed that stopping the succession at the subclimax could be easily accomplished through human intervention. Hofmann was certain that the wind could not have been responsible for seeding areas more than about a hundred feet away from any seed trees. This assumption, like the ability of stored seed to germinate, was pivotally important for the verity of his theory. In his observations of burned areas, he interpreted each site's pattern of seedling growth as showing seed tree effects only very close to those trees, while seedlings farther away he attributed entirely to seed stored in the duff. His conclusions were always definitive, as in his statement that the "peculiar distribution of the reproduction . . . shows very definitely that the green timber remaining after the fire has had very little influence on the general occurrence of the Douglas fir reproduction over the burn."[10] His interpretations of such observations are notable for their reluctance to consider any other possible cause for the observed patterns of seedling growth, such as differences in soil, topography, or animal activity. Hofmann stated that all the factors that might lead to the replacement of Douglas fir with the climax species were "within the control of man, and it is on [these factors] . . . that the scientific management of Douglas fir must be based in order to keep the Pacific Northwest region under continuous natural production of this most important species."[11]

By the mid-1920s, lands that had been prepared after clear-cutting in the late 1910s according to the hands-off prescription of J. V. Hofmann should have begun to show the first signs of regeneration. In a great number of these sites, however, there was little evidence of any regeneration of the Douglas fir; instead there was growth of brushy and noncommercial species, or worse yet, landslides and other devastation of the landscape. Some foresters, publicly or privately employed, never had accepted the seed-storage theory in the first place and were now pointing to these failures in reforestation as refutation of the theory. These critics included such well-known figures as E. T. Allen, a former Forest Service regional forester who had become the executive director of the Western Forestry and Conservation Association, a lumber industry lobbying and public relations concern. The combined pressure of public and private foresters' comments, the obvious failures of cutover sites, and finally Hofmann's 1924 resignation from his post at the Wind River Experiment Station all together led to the long-overdue reassessment of Hofmann's theory.[12]

Bob Marshall examines the forest floor as a United States Forest Service research forester at Montana's Northern Rocky Mountain Forest Experiment Station (1928). Marshall's research as a Forest Service scientist would influence his later dissertation work on the soil conditions seedlings needed for optimal growth. Photograph 229350; Records of the Forest Service, RG 95; National Archives at College Park, MD.

The project of retesting the elements of the seed-storage theory fell to Hofmann's replacement as director of the Wind River Experiment Station, Leo Isaac. With a true passion for fieldwork, Isaac said that "working out the mechanical details and the field work are simply pleasure for me," but, in contrast to some foresters' constant publication, "the literary end of the game is my stumbling block." Isaac maintained a rigorously scientific mind-set in his research while simultaneously working toward keeping the forests under his watch "safe from the destructive hand of the logger" when possible.[13] Indeed, Isaac wished for intact forests preserved from lumber activity, hoping that before it was too late, "a few of

my dreams will be realized—a few remnants of virgin forests will be preserved
along our highways and a few miles of our Oregon coastline will be preserved
in its natural state."[14] While he was not publicly vocal about the value of wild
forest, it is clear that privately he worried about the unchecked growth of the
lumber industry in the region. Although most foresters were ready to reexamine
Hofmann's theory, there was still a large contingent of adherents, especially within
the region's lumber companies. Isaac's project, then, would be vital not only to
further scientific knowledge of the Douglas fir forest but also to inform methods
of forest regeneration on thousands of acres of cutover lands.

In 1925, Isaac began experiments designed to address both of the most
problematic areas of Hofmann's work: the assertion that the seed could remain
viable for the long term in the soil, and the assertion that long-distance wind-
borne dissipation of the seed was uncommon. The first experiment to produce
results was his exploration of the flight patterns of Douglas fir seeds. To replicate
the height at which the trees' cones released their seeds, Isaac used a custom-
built kite, from which seed-dispersal units made from oatmeal canisters were
suspended. When the kite attained the proper altitude, a string tripped the doors
of the canisters, letting the seed fly out. While the oatmeal canister did not,
of course, mimic the mechanism by which the cones naturally dispersed seed,
it did replicate the wind and altitude conditions of the seed's release. Because
the trials were done in winter, the researchers could easily find the seed after it
had landed on the snow. Isaac found that the seeds traveled much farther than
Hofmann had assumed, often as much as a quarter mile from the release site.
Isaac performed his experiments in the landscape in question, and his kite allowed
a near approximation of the wind conditions in the top of a Douglas fir. Thus,
as near as possible, this experiment directly addressed Hofmann's theory without
necessitating any great assumptions.[15] Isaac also began work on experiments that
would refute Hofmann's assertion that Douglas fir seed could remain viable in
the duff for several years. Isaac buried seeds in the duff, not only of Douglas fir
but of seven other Northwest forest species as well. To avoid the rodent problems
that had thwarted previous seed-germination experiments, Isaac placed the seeds,
buried in duff, in cages, and then buried the cages in the forest in an approximation
of natural conditions. No germination was found after the first or subsequent
years.[16] Isaac's double-edged refutation of Hofmann's theory convinced many
who had been hanging on to it because of the appeal of its utility and convenience
for regeneration.[17]

The Ecological Education of Bob Marshall

Leo Isaac had several assistants in his work on seed flight at Wind River, young professionals in their first years out of forestry school. These assistant foresters saw firsthand the value of ecological thinking in forestry, and its potential to revolutionize the methods and aims of forest management. The work of an assistant forester in those years entailed the grunt work of field experimentation: the heavy lifting and mind-numbing tabulation of data behind every experimental result. The experience gave assistants exposure to the daily practice of forest science and acculturated them to employment in the US Forest Service. One of these assistants was an indefatigable, intellectually curious, and ambitious young New Yorker named Bob Marshall. When Marshall was at Wind River, it was still early in his career, and Isaac was just beginning the seed-dispersal studies that would eventually discredit Hofmann. Being present during the initial stages of this project allowed Marshall a view of Isaac's creative process in developing the experiments' design and parameters. Isaac would later reminisce that he had "yet to find a companion in work more willing, kindly, and fair" than Bob Marshall had been during his years as Isaac's assistant.[18] Marshall would become a divisive figure in 1930s forestry, and an important advocate for wilderness preservation. His values, education, and background guided his visions for the futures of both professional forestry and wild forests.

Marshall's experiences assisting Isaac in the Cascades were his introduction to forest ecology, but he had already spent much time in the forest before then. While he was raised in the midst of great wealth, he was not simply the product of an ordinary privileged New York City upbringing. Louis Marshall was a well-known figure in the intellectual and political life of New York City in the early decades of the century. A prominent corporate and constitutional lawyer, the elder Marshall also became known for mediating and arbitrating trade union strikes and for generously donating time and money to civil rights causes, and by the mid-1910s he had become one of the most prominent Jewish leaders in the city.[19] He was also deeply interested in conservation of the forests and mountains of the Adirondack region around the Marshall family's summer home, Knollwood, on Lower Saranac Lake. The elder Marshall was also credited with inserting the phrase "forever wild" in Article 14 of the New York State Constitution, a provision that requires that a constitutional amendment be passed if the state wishes to develop any part of the Adirondacks. As discussed in a previous chapter, the state of New York had provided for the establishment of a forestry school in 1898, acknowledging the need for trained experts to manage the state's massive upstate woodlands as well as the forests of the rest of the nation. The school, located at Cornell, had been shut down after just five years, as a result of political fights over appropriations. In 1911, Louis Marshall was part of a group of conservationists and lovers of the Adirondacks that had succeeded in pressuring the state legislature to fund a new forestry school to replace the one lost earlier. The new version of the New York State College of Forestry was established at Syracuse University the following year. It distinguished itself by emphasizing the development of expertise in specific scientific fields and focus areas, rather than simply offering a general forestry curriculum for all students.[20]

The young Bob Marshall spent his school days studying at the progressive and cosmopolitan Ethical Culture School on Manhattan's Upper West Side, and his summer days climbing peaks and exploring forests in the remotest sections of the Adirondacks. Upon his graduation, he determined to be a professional forester, a career that would meld the Marshall family values of a love of the outdoors, intellectualism, and a passion for public service.[21] Bob Marshall attended the resurrected New York School of Forestry in Syracuse, which his father had helped ensure was one of the larger forestry schools in the country by the time of his matriculation. Syracuse fostered Marshall's interests in both the ecological study and recreational management of forests and helped him develop a critical eye toward the processes and habits of scientists. Even during his college years, Bob Marshall was thinking about the difficulty of reconciling the multiple irreconcilable demands that Americans placed on their forests, mulling over the balance between economic and recreational use of the Adirondack forests.[22] The elder Marshall had instilled in his son a firm set of left-wing political views, a fervent need to work in behalf of those views, and a compulsion to speak up for his beliefs. Throughout his life he would work to meet his father's values as he developed an approach to forestry that was both ecological in its scientific approach and radical in its political approach.

Upon graduation, Marshall was hired into the research branch of the Forest Service and moved west to work at a series of Forest Service experiment stations. He quickly made a name for himself, both for being an intelligent and diligent forester and for being a cheerful misfit in forester culture.[23] This dichotomy can be seen, for example, in the two papers he simultaneously published about forest precipitation effects, one very serious and one an elaborate joke. Lighthearted in many parts of his life, he habitually lampooned professional customs and pretentions. For example, his attitude toward the scientific establishment and the practice of science was well demonstrated in the series of comical scientific "experiments" he conducted, and even occasionally wrote up and managed to publish. One of his best, published in *The Nation* in 1927, is entitled "Precipitation and Presidents." In this paper, Marshall made the argument that rainfall, through influencing crop yields and hence economic health, indirectly affected voters' actions. As the region grew less dependent on agriculture, the correlation between precipitation and voting trends lessened. "Precipitation and Presidents" was a loving, pitch-perfect parody of science. Marshall mimicked the tone and terminology of scientific writing but presented the argument lightheartedly. While explaining his methods he wrote that "we are blessed in the United States with a national baby show and a national beauty contest, a national flag and a national flower . . . almost everything seems to be nationalized except the weather. That still maintains a complete regional individuality."[24] Indeed, the concept of climatic determinism was then in style, and "Precipitation and Presidents" was deadpan enough that it could almost be misconstrued as seriously intended. Ellsworth Huntington, a Yale geographer who had served as president for both the Ecological Society of America and the Association of American Geographers, was one of the best-known proponents of the theory of climatic determinism. In 1912 he had published a study using sequoia tree rings to correlate historical events with climate using the same methods that Marshall had used, but surveying a far wider swath of history and using a wider range of ecological data.[25] The shaky logic and broad conclusions of Huntington's widely heralded work then served as inspiration for Marshall's spoof.

Marshall crafted this parody concurrently with his own serious research on forest climate. He had conducted a study of the ecology of precipitation in western forests while on the job at the Northern Rocky Mountain Forest Experiment Station earlier in the same year. He presented these data first to the Northern Rocky Mountain Section of the Society of American Foresters and then published it as "Influence on Precipitation Cycles in Forestry" in the April 1927 issue of the *Journal of Forestry*. Marshall pointed out that "with such surprising unanimity among ecologists, it seems strange that foresters for the most part have neglected this clue [radial growth of trees] to past climate."[26] Long-term patterns and cycles in climate were of interest to ecology but also aided in foresters' management decisions about firefighting and reforestation and were valuable for predicting future growth rates of forests. The paper, though written in a slightly looser style than most of its type, engaged in serious data analysis and persuasive conclusions. The allusions to Huntington's style were absent from this paper, and all conclusions remained firmly within the realm of ecological thought. Marshall therefore demonstrated that he could approach a single topic both seriously and comically, the simultaneity of the two papers creating an amusement only very few would have noticed. While others took the profession seriously, he played with its boundaries and habits, even mocking venerable scholars like Huntington. These two papers, taken together, can be seen as an encapsulation of Marshall's flippant yet brainy professional attitude. Years later Leo Isaac remembered the Bob Marshall of this era as gregarious and energetic, but also somewhat detached, not particularly wedded to forestry or its conventions.[27]

In 1926, while recuperation from an ulcer forced him to take temporary leave of his position with the Forest Service, Marshall wrote his supervisor about the possibility of taking a longer leave of absence to earn a doctorate. The interwar decades saw a dramatic increase in the status and prevalence of professional degrees among the ranks of many professions, and the Forest Service was no different. In the hope of increasing the numbers of employees holding advanced degrees, the Forest Service developed an initiative to increase the professionalization of those already within its ranks. As part of the general push toward improving research within the agency, employees were allowed and encouraged to take leaves of absence from the profession to return to universities for doctorates in their fields of specialization. They pursued research in a variety of scientific fields, including wood chemistry and range management, but most chose either forest ecology or silviculture. Marshall's supervisor urged him to postpone his leave, in order to better acquaint himself with the realities of forestry and the habits of fieldwork. "Although you are strong on the essential qualities desirable in a novice either in research or administration," the supervisor wrote, "you are very weak in the mechanics of field work and field living."[28] Two years later, a seasoned Marshall knew more about the mechanics of fieldwork, to be sure. However, his time with Leo Isaac and other experiment station scientists also helped him better understand ecology, both as a research approach and as a profession. Bob Marshall was one of thirty-two Forest Service employees to take advantage of that program when he left for Johns Hopkins University.[29]

At Hopkins, Marshall studied under Burton Livingston, an ecologist who had done extensive ecological fieldwork in the first decade of the twentieth century as part of the founding generation of American ecologists. Many other labs would have afforded him reason to continue the habit of fieldwork but would not have provided the intellectual community and theoretical innovation of the Livingston lab. By the 1910s Livingston had incorporated plant physiology into his research, but his research group remained focused on ecological processes, specifically the relation of plants to their abiotic surroundings. By the time Marshall arrived, Livingston had perfected a series of instruments that allowed accurate field measurements of the physiological processes of plants while they remained living in their original ecological settings. Livingston had also become witheringly critical of the holistic underpinnings of Clementsian ecology. Livingston and his research group were an active part of ecological discourse even though their work centered in the greenhouse and the laboratory rather than the field. Marshall, who had grown comfortable with experimental ecology at Wind River, was a natural fit. Burton E. Livingston was a highly regarded plant ecologist and a driving force in the development of tools and techniques for field experiments. He was the director of the Laboratory of Plant Physiology at Johns Hopkins, during an era when plant physiology was a part of ecology and research took place mostly in the field.[30] Livingston was instrumental in the professionalization of plant ecology by training a generation of scientists and promoting and effecting the shift toward quantitative methods of measuring dynamic ecological phenomena.[31]

Livingston demanded rigor in ecological practice, working to bring field experimentation in line with the underlying theories of ecological dynamics. Trained as a plant physiologist at the University of Chicago, Livingston would have a lifelong focus on the dynamic interchange between plants and the air, light, and soils that support them. This was a physiological ecology;[32] ecology as the study of species locked in continual interaction with the richness and paucity of their environments. Livingston's lab stayed on the cutting edge of ecology, studying little-understood natural phenomena by using entirely new techniques and instruments. Livingston's students did not simply collect and observe in the manner of natural historians. The new ecological questions with which he was concerned differed from older questions not just in topic but in the object of scrutiny. Previously, most ecologists had studied the same discrete building blocks of the world that the generations of natural historians before them had—the plants, animals, soils, and waters that make up its constituent parts—and simply rearranged these building blocks in different ways.

Marshall's decision to give up western forest fieldwork for a Baltimore lab suggests the seriousness of his ecological interests at this point in his career. Marshall's doctoral research, while it made use of his background in forestry, was still an unalloyed product of the ecological sensibilities and theoretical stance of the Livingston lab. Marshall studied the relationship between plants and their water source, and the physiological process of water uptake from the soil into the plant. He focused especially on wilting in commercially important conifer species that he brought to the Hopkins campus greenhouses. In the 1920s water uptake had often been used as a measure of physiological activity in a plant, with researchers measuring fluctuations in soil moisture to estimate water uptake by plant roots. They had used that statistic to predict wilting, thus indirectly estimating drought resistance in natural settings. However, inferring physiological processes through observed fluctuations in soil moisture was an unreliable way to determine the water uptake process.[33] Livingston led a new approach to understanding water uptake by "turn[ing] attention to the dynamic water-supplying power of the soil—in other words, the rate at which water was available to the absorbing surface of the plant."[34] While Marshall's conclusions had ramifications for determining the drought resistance of natural and artificial reforestation of these economically important species, he did not develop such recommendations in the published thesis. The species Marshall elected to work with were of significant interest for the Forest Service because of both their economic value and their widespread occurrence on national forests. Indeed, Marshall could have chosen a species whose wilting could be seen by the eye, simply for its ease of use in the lab, as others in the Livingston lab did.[35] Marshall determined that satisfying the ecological questions about his chosen class of plant trumped the difficulty of working with them in the lab.[36] Furthermore, after the experience of working with Isaac on the problem of the Douglas fir, Marshall knew well the importance of having scientifically rigorous work to rely on for creating successful reforestation plans.

With his choice of species and topic, Marshall contributed to federal forestry while still doing research that was inarguably ecological. This research supplied answers to questions raised by the management needs of forestry but was also part of the vanguard of experimental ecology. His doctoral thesis introduction specified that "the author wishes to acknowledge indebtedness to the pioneering publications"[37] of Raphael Zon and Carlos G. Bates for lighting the way in combining ecology and forestry. Bates and Zon had produced *Research Methods in the Study of Forest Environment*, which made a strong case for the importance of quantitative ecological investigations in the advancement of innovations in forestry. Bates and Zon's book had included significant discussion of Livingston's techniques, instruments, and experiments.[38] Indeed, *Research Methods in the Study of Forest Environment* was remarkable for its time, introducing a level of ecological rigor into federal forestry and reaffirming the scientific stance of the federal forest experiment stations.[39] His research species included Norway spruce, a European species that was rapidly growing in favor for industrial reforestation in North America, and the "Norway pine," or red pine, a hardy pine that (despite its misleading common name) is native to the forests of northeastern North America. The other species in the study included Rocky Mountain pine species;[40] two species popular in southeastern pulpwood plantations;[41] and two natives of the eastern boreal forests.[42] Completing the list of twelve were four major species of the Pacific Northwest: the Douglas fir, the shade-tolerant western hemlock and western red cedar, and the "lowland white fir," better known as the grand fir and strongly identified with the Olympic Peninsula's lowland rain forests.[43]

Marshall admitted that his study of conifer wilting was "broad rather than deep."[44] He aimed to provide guidance on seedling characteristics for a number of economically important species, rather than exhaustive study of a single test species. He cautioned that his "study was undertaken mainly as an attempt to develop useful methods and as a sort of reconnaissance survey of possibilities, rather than to secure numerical results that might, in themselves, be reliable as representing the several species and the several experimental treatments."[45] While the results achieved "are only suggestive and indicative at best" of fully quantitative descriptions of the physiology of wilting, "they do show possibilities and probabilities" and hence "may be useful in planning more thorough experimentation in this important field of seedling physiology and forest ecology."[46] As Leo Isaac had proved, research into seedling growth could help correct mistakes in earlier decisions about reforestation and industrial forest management. The lack of knowledge surrounding regeneration of important tree species affected many of the country's logging areas, not just the Douglas fir. The US Forest Service had been encouraging research into natural and artificial reforestation through germination and seedling studies, as well as larger-scale management investigations, since the days of Zon's Research Division. Conifer species Marshall studied included many of interest to national forest management, and he sourced much of his seed from the Forest Service.[47] The species he chose included merchantable timber or pulpwood species, competitors with merchantable species, and fast-growing exotic replacements for native species.[48]

The Zon Coterie: Ecology, American Forests, and Social Utility

While he researched forest ecology, Marshall also advocated for radical change for American public lands and for government at large. In the years before he arrived at Hopkins, Marshall had written several articles about the possibility of socializing American forests. He wrote about forest policy to try to steer the future of America's forests, to save the wilderness experience. He had been raised in the midst of political activism and reflexively spoke out about what he believed was wrong. Deeply driven by his political beliefs, he intended much of his written work to steer the government, professional societies, or other bureaucratic entities in a particular direction or toward a specific issue. He did that simply to enable future exploration, though, not as an end unto itself. This work was separate, however, from the straightforward forestry he had done at Syracuse and in the Forest Service. It was likewise separate from the hiking and canoeing expeditions he had done in his free time. Marshall was driven by conscience to his work in forest policy, but driven by love of nature to his work in science. He had entered forestry because he loved the forest and wanted a career that would allow him to spend time outdoors. His conscience and temperament would lead him to desk jobs, but temperamentally, he was definitely a scientist and an adventurer, not a bureaucrat.

It was during his time at Hopkins that Marshall first encountered the individuals with whom he would form a small coterie of radical, ecologically oriented foresters. The group coalesced around Gifford Pinchot but was intellectually driven by the scientific and political thoughts of Raphael Zon. Marshall's status as Livingston's PhD student proved his credentials to the highest levels of ecological forestry. This gave him credibility in a way his family wealth or his published articles never could. While the depth and sincerity of Marshall's politics were never in doubt, his intellectual standing as a scientist might have been questionable before his time at Hopkins. By sacrificing his carefree western lifestyle for the hard work of basic science, Marshall proved himself more capable and responsible than he might otherwise have seemed. Further, in the time he was at Johns Hopkins, he made his first real connections with an elite group of current and former federal foresters who were at least as disgruntled with the state of federal forestry as he was. His return to the halls of academia, and to the intellectual life of the East Coast, seems to have partly enabled his entrée into this group.

Early in January 1929, Marshall wrote a letter to George Ahern, an outspoken elder of professional forestry. Ahern, then the president of the Washington chapter of the Society of American Foresters, was a career army officer and friend of Pinchot whose career high point had been as chief forester in the territorial Philippines.[49] The previous month, Ahern and Pinchot had appeared at a Society of American Foresters meeting in New York to promote Ahern's recently published *Deforested America*. When he returned from the meeting, Marshall immediately read the book, a chronicle of the increasing impact of industrial lumbering and a call for federal control, which had been published with Pinchot's money. Marshall appreciated Ahern's call to action, writing that "for 3 ½ years out in District 1 I have been arguing feebly and futily [*sic*] with my forester friends upon the very proposition which you have so ably presented, and you can have no idea how joyful I am at this outspoken voicing of the sentiment by a leading forester." He continued, "I take it that you and Mr. Pinchot are contemplating some organized effort to put across the ideas which you have raised." Neither the professional Society of American Foresters nor the populist American Forestry Association could be the proper venue for their work, he said, "since the large number of lumbermen and timorous conservatives which both contain make it impossible for them to do more than pass resolutions and pursue vacilating [*sic*] courses. A dynamic new organization composed exclusively of those who favor vigorous government control seems imperative. I would hesitate to write about this matter, appreciating fully that you and Mr. Pinchot do not require my kickshaw advice, were it not for the fact that such a new organization will need money *immediately*, and due to certain accidental influential connections I think that I might be able to help raise some of it." While that offer of funding was sure to gain their attention, his writings proved he was more than a mere dilettante. He enclosed a draft of a highly laudatory review he had written of *Deforested America* and dropped mention that he had "already had an article printed in the *Nation*, and think I could get this one in. It seemed to me that everyone who feels keenly about the idea you raised should do his best to give it wide publicity."[50] Marshall concluded the letter with a summary of his career, acknowledging that "it is always disconcerting to receive a long letter from a person about whom one knows nothing."[51]

Ahern responded quickly, recommending the young forester go forward with the article he planned and expand his critique of the current federal forest system. This became the start of an association that would last for years. Soon after his initial contact with Ahern, Marshall received a call from Gifford Pinchot, inviting him to travel from Baltimore to the Pinchot home in Washington, DC, for an informal meeting with other concerned foresters. The Pinchot home had been a famous site for discussions of forestry and forest policy since his earliest days in the capital. After Pinchot left the Forest Service the meetings continued, as he brought together his friends and allies to discuss developments in the field. Strategizing from behind the scenes, Pinchot continued to exert influence over policy. At this meeting the young Marshall got his first chance to mix with the esteemed elders in the profession, including Ahern, Pinchot, and Zon. Marshall injected fresh ideas and new energy into this group of like-minded foresters. Idealistic and anti-industrial, this group was motivated to slow forest destruction on public lands. They also shared a common belief that their profession had lost the focus it had once had. While all of these men were individually critical of forestry and forest policy, through their association they would develop a larger and more effective voice. This coterie of radical foresters, which viewed Zon as their intellectual heart, would develop a powerful critique of their profession and its views of the lumber industry.[52]

For Marshall, 1929 was a year for branching out in new professional directions. In the spring he developed his intellectual and political views with the supportive coterie of radical foresters, and during the summer he left the sweltering city behind to explore the forested wilds of Alaska. He had resolved to make the most of the time between the two years of his doctoral program by traveling in some of the wildest, least industrial areas of the continent. While other doctoral students worked in labs or attended summer field sites, Marshall had both the money and the mind-set to set out on a sketchily planned adventure in the High Arctic of Alaska.[53] Marshall's new ally, Gifford Pinchot, who had shunned staid placidity his entire life, was an inspiration for both Marshall's 1930 journey and the book that chronicled it. At the end of March 1929, as Marshall wrapped up his studies and prepared for Alaska, Pinchot left on his own adventure in the South Seas. A yacht voyage from the East Coast through the Panama Canal to the Galápagos and Tahiti, Pinchot's trip, like Marshall's, had multiple goals. While on the one hand it was a long-planned adventure to a part of the world that held great allure, on the other hand it was a formal collecting expedition to a region still little studied by biologists. With representatives of the US Biological Survey and the Philadelphia Academy of Natural History on board, the Pinchot expedition collected bird specimens for the museums, discovered and named several new species of land snail, and procured a live Galápagos turtle for the Philadelphia Zoo. It was also a grand adventure for the Pinchots, a break from the norm after Pinchot's defeat in his campaign for reelection. Pinchot produced a lavishly illustrated narrative of the journey, written for an audience who desired not just a dry account of natural history but a tale of adventure that also depicted the day-to-day activities of a collecting expedition and the exploits of one of Pennsylvania's most prominent political figures. The aim was a scientific adventure story, because, as Pinchot wrote, "adventure seasoned with science is the very best kind."[54] Pinchot returned from the voyage in late December of 1929 and was in the midst of writing *To the South Seas* when Marshall was planning his own adventure seasoned with science.

Isaac's work in seed dispersal had inspired Marshall to think about how Douglas fir forests colonized treeless areas, and Marshall pursued that same topic in his ecological inquiries into the northern treeline in Alaska. The forests of Alaska give way to treeless tundra at the most northerly latitudes, but the reasons for the specific location of the treeline were not certain. Most scientists thought that, as with the altitude-delineated treelines on mountains, the northern treeline was governed by the harshness of the climate. As the weather was colder closer to the North Pole, many assumed that at a certain point trees were simply unable to grow because of the harsh climatic conditions. While this seemed reasonable by analogy with the better-understood alpine treelines, no ecologists had yet conducted any real scientific inquiry into the phenomenon. Marshall suggested that perhaps "the northern timber line in Alaska is not the result of unfavorable environment for tree growth, but simply of the fact that there has not been time since the last ice sheet receded for the forest to migrate further north. According to my theory, the spruce stands eventually will extend to the Arctic Divide and cross over into the sheltered valleys north of the divide."[55] As written by a Jewish New Yorker in love with the high country of the American West and drawn to the harsh landscapes of the Arctic, these words have a certain poignant significance. The ecological question of whether a species native to the

great forests of the Adirondacks could hope to thrive in the far north may be seen
as having a poetic, if purely metaphorical, resonance within the larger biography
of Bob Marshall.[56]

There have always been scientists driven by their desire to adventure, to
examine new and unique specimens, to lay eyes on beautiful or rugged landscapes.
Marshall's Arctic science was in this vein: he wanted to experience these remote
areas, as well as find out what they were like ecologically. His chosen site, near
the Arctic Divide, was so remote that he reached it only after "a couple weeks
of backpacking" from the nearest town.[57] Marshall was drawn to this region
that offered, at the time, a forested world virtually untouched by industry and
convenience.[58] Wherever he found himself on his Alaskan hikes and travels,
Marshall sampled and surveyed the forests to add to his store of data on the topic
of the northern timberline. While accompanying local hunters on a November
dogsledding trip quite close to the timberline he spent some spare time surveying
a stand near camp. "With much pressure and at the price of numerous shivers and
some profanity I bored nine different trees. . . . I just had time before darkness
to cruise a quarter-acre sample plot in the heavy timber surrounding us." While
boring trees within a stand gave exact ages of those individual trees, it was
difficult to gain reliable generalizations from such samples. The timber cruise, a
staple of the forester's profession, gave a general sense of the forest by sampling
all the trees within a predetermined plot to gain statistical figures for the size and
number of trees per acre within the forest as a whole.[59] Marshall was heartened
by the results from these efforts, writing that "the ages and distribution of the
timber gave welcome confirmation to my theory on the advance of the northern
timber line. . . . It may be disconcerting to those who hold the commonly accepted
theory of a stunted northern timber line to learn that just a mile and a half away
[from the current timber line] were trees 18 inches in diameter at breast height."[60]

Marshall designed an experiment to test this theory that was similar to the
test plantings that Leo Isaac had placed around the Wind River, to test how well
tree species from around the United States adapted to local conditions. Marshall
suspected that it should be possible to test his theory "simply by sowing seed to
extend the timber line far north of where it now is. So I collected spruce cones
about four miles south of the last timber, extracted the seeds, and sowed them on
two plots of ground on Grizzly Creek, twelve miles north of the present timber
line." The seed planting was done within parameters of experimental design: "One
plot was covered with the natural vegetation. . . . On the other I scraped away the
vegetation and sowed the seeds directly on the black soil." Marshall concluded by
observing that if any of the sown seeds were to germinate and become established,
the growth of the planted spruce there would mean that "this experiment

constituted an advancement of the timber line of about 3,000 years according to my estimate of spruce-migration rates." Along with the experimental plantings, Marshall established a protocol of extensive climate monitoring at a scientific station he set up on the outskirts of Wiseman. The value of this sampling work for Marshall's theory of the northern treeline would be to provide reliable base data for the area, information he could refer to as he worked toward conclusions about the continual effects of climate on species distribution.[61]

During his stay of two months in summer 1929, Marshall gathered little conclusive evidence about the northern treeline, admitting that on that visit, "I cannot say that I learned very much either about tree growth or timberline."[62] However, he had become interested enough in Alaska to develop plans for a return. After completing his doctoral work Marshall would return to Alaska in the summer of 1930 to continue his research on the treeline as well as explore more of the country. In a letter to his friends Gerry and Lily Kempff, written just before he left for his second Alaska trip, he detailed his scientific plans. Not only was he going to continue the ecological studies he had begun the previous summer, he would also "make some studies of the physiology of the transpiration in this far northern region which fascinates me very much." In temperate zones, plants lose water vapor during the day as a side effect of the process of transpiration, but at night the plants have a net gain of water vapor "and the water balance is restored." Were it not so, the plant would eventually wilt from loss of water. However, Marshall went on, "what in the world happens in a section where one has to wait four months for night to come around? According to all that is just and holy, the plants should go into permanent wilting long before night arrives. And yet the fact remains that they dont [sic]. It should be quite exciting to find out why." Marshall's excitement extended to studies of other aspects of Arctic forest ecology as well. He also mentioned his other plans, including collecting specimens for various Hopkins laboratories, doing "a great deal of rather heavy reading," and "some more of that socially useless but egotistically and aesthetically fastinating pasttime [sic] of exploring country with stupendous scenery which has never before been visited by man."[63]

Just before he left for Alaska in July of 1930, Marshall's Forest Service friend Ward Shepard told him that "like all modern explorers I know you are fully equipped with impeccable scientific motives and know how and when to use them; but pray feel free to relax and tell us from time to time of your most atavistic emotions and primal impulses. We shall require no tangible economic, ethnic, or geographic returns."[64] While ecological research still excited him, Marshall did indeed expand his focus from the narrow ecological questions he had described to the Kempffs. The 1930 trip involved a more concerted effort to socialize with small-town Alaskans while still maintaining a distant, almost scientific stance toward those he got to know. Marshall's central project during this second trip would be writing a narrative of life and culture in the small and, as he called it, "preindustrial" community of Wiseman, Alaska. On his first visit he would come to love not only the wilds of Alaska but also the people and the community, and with this book he hoped to capture their world before it disappeared. "I think the study should be made, I don't know anyone else who is likely to do it before it is too late, I know I will have a wonderful time out of it, [and] I have an offer for publication for a book on the subject." Perhaps self-conscious that most people, upon receiving a doctorate, moved directly into the next stage of professional life in their field, he added, "I feel before settling down too narrowly that I should broaden myself out by some study of the social sciences, [and] I crave a little additional adventure before settling down to staid placidity of middle age, and so, of course, I am going to go."[65]

While Gifford Pinchot gathered rare specimens and described animal behavior on his trip to the Galápagos, Bob Marshall recorded dendrographic data and planted experimental plots on his trip to the Arctic. However, both Pinchot in the Galápagos and Marshall in Alaska conducted these scientific investigations according to Pinchot's declaration that "adventure seasoned with science is the very best kind."[66] For both, the trips were self-financed and self-directed and were conducted primarily to indulge their interests in the unusual or interesting characteristics of the sites and their ecology. The questions the investigations were tailored to solve were those that attracted the curiosity of the investigators, not those that would speak to the economic needs of society. This sort of science could be important groundwork for more directed research, but it did not try to justify itself in that way.[67] On the first page of *Arctic Village* Marshall introduced himself as "a forester and plant physiologist"[68] whose view of Wiseman would be that of a sympathetic outsider. Although he acknowledged there was plenty of adventure during his time in Alaska, and plenty of science too, writing about such was not the purpose of the book. Instead, he wrote, "I am writing this book with the purpose of painting a complete picture of the civilization of whites and Eskimos" in that region, "to describe in an objective manner the unusual civilization" of Wiseman and the "independent, exciting and friendly life of the Arctic frontier."[69] Above all, *Arctic Village* reads as a chronicle of Marshall's own explorations of the meaning of life and the foundations of happiness. In 1933, Marshall wrote that after returning from his 1930 trip he "frankly acknowledged that the justification for exploration in modern times must be found primarily in what it contributes to the personal happiness of the explorer rather than in what it may add to the well-being of the human race."[70] During his time in Alaska, he struck a balance between his study of ecology, his desire for adventure, and his interests in culture. He would continue to seek that balance for the rest of his career.

The "Letter to Foresters" and the Beginnings of the Zon Coterie

By the start of 1930, Pinchot, Ahern, and Zon had met and corresponded enough with the young Marshall to consider him a professional equal. The four of them, led by Zon's strong intellect and passion, formed a circle constantly corresponding with one another about radical reform of federal forest policy. Along with a revolving group of other politically radical, ecologically inclined foresters, Zon's coterie developed strongly critical positions on both forest science and forest management. The first tangible result of this alliance was a collaborative document urging foresters to embrace a wider role for the profession, adopting a socially conscious stance toward matters of forest use and ownership. In early 1930, Pinchot was beginning his campaign for a second term as governor, splitting his time between Washington, DC, and Pennsylvania. Although in letters and discussions he often weighed in on forestry matters, he no longer had any official role in the profession. After Pinchot's departure from the Forest Service, Zon had suffered waning influence and failing allegiances within the agency. His abrasive manner, which friends characterized as "sniping and scolding," and a tendency toward "carrying criticism too far"[71] on forestry matters, coupled with an outspoken leftist political stance, impaired his career during the conservative early 1920s. In 1923 he relocated to Minnesota to direct the Forest Service's Lake States Forest Experiment Station. While not a demotion per se, Zon's move semiexiled him both managerially and geographically from the power center of federal forestry. And Ahern, while based in Washington, had worked in forest management overseas and never been part of the inner circles of federal forestry. Pinchot, Zon, and Ahern thus were all monitoring the Washington activities of the Forest Service at some remove.

Sensing that federal forest management was increasingly catering to lumber interests, Pinchot and Zon concocted a plan to confront their colleagues and demand change. Pinchot invited Marshall to his home, where they collaborated to outline a critique of professional foresters. While the points of criticism originated with Zon, he declined to write the letter himself. Worried about causing further damage to his career, which had already suffered from his transfer to Minnesota, Zon wrote that the brave young Marshall would be less likely to pull punches and more able to write in the proper tone of excoriation.[72] The document, titled simply "A Letter to Foresters," was written by Marshall with help from his friend Ward Shepard. It was signed by Zon, Pinchot, Marshall, and Ahern, along with Ward Shepard and two other federal foresters, E. N. Munns and W. N. Sparhawk. First mailed privately to the member list of the Society of American Foresters, the letter generated much discussion and was subsequently printed in the *Journal of Forestry*. The "Letter to Foresters" pointed out that the nation's forests, particularly the private forests, were in the midst of a period of immense destruction, and foresters had a duty to stop it. It would be "a moral tragedy," the letter stated, if "the foresters of America will accept that destruction and by silence condone it."[73] The social role of foresters was a fundamental part of the profession, the Zon coterie wrote, and foresters' "failure to grapple with the problem of forest destruction threatens the usefulness of our profession."[74] American forestry was "born with high ideals and great purposes. It has fought many a bitter fight against heavy odds. It has won magnificent victories. From the very first its guiding spirit has been that of public service," but, they lamented, that spirit had recently decayed.[75]

They warned that if the profession ignored questions of the larger social good in favor of solving industrial forest management concerns, it would be going down a dark path. The letter observed that although foresters widely acknowledged this devastation, few of them were engaged with stopping the damage. Professional foresters, they wrote, should be working for two things: first, public regulation of private forest practices, and second, "a greatly increased program of public forests." If it was accepted that the ultimate goal of the profession was "the safety and prosperity of our country," then the importance of keeping the nation's forests safe from harm should be evident. Further, not just forestry, but forests, were in peril. If foresters did not renew their commitment to public good and keep industry at arm's length, the nation's remaining intact forestlands would be destroyed for profit. "The future of our forests," they warned, "of our forest industries, of organized forestry agencies, of education in forestry, and of the profession itself is all dependent on stopping forest destruction." The public regulation of private forests was important for three reasons. First, unregulated logging of private forests could affect entire watersheds, creating damage to public water supplies, agriculture, and other properties. Second, abandoned cutover private forestlands were rapidly reverting to public ownership when private owners decided to stop bearing the financial burden of their ownership. And finally, the future welfare of both the forestry profession and the nation's forest industries rose and fell with the fate of the private forest. The letter noted that "the silvicultural basis for control" of logging had been determined by Forest Service research, but that the agency had no resolve to apply that control to industry practices in most of the country. The program of basic research that Zon had spearheaded was in peril due to the loss of natural forests. Forest destruction would damage the progress of science as well, they pointed out, asking, "To what end a great program of forest research if the forests to which it should minister are to be destroyed?"[76] If natural forests dwindled into nothing, Forest Service programs would be reduced to catering to industrial needs. The research that would remain would be only that which met the needs of the forest industry—examinations of reforestation, fire control, and insect control.

Most of the Zon coterie members were also prolific writers of policy articles and editorials, but they garnered more attention by banding together to create this single document. This multiauthored call to arms gained more notice than solo articles written by even the venerable Pinchot. The Zon coterie framed their defense of the profession as a moral struggle against the corrosive force of industry, and such a message was more powerful if it came from a chorus of voices. Denying the social role for the forester was a moral failing for the profession, they said, because "failure to meet responsibility is implacably punished by spiritual decay.... We must cleanse our minds of apathy and doubt; and through a rebirth of faith in forestry and a reawakening of all our moral and mental energies, we must set the forestry movement on the path to its goal."[77] The forester was a caretaker of the Earth, and compromising that goal was a serious breach of ethics.

When the letter was published in the *Journal of Forestry*, it appeared alongside several responses from foresters in academia and industry. These respondents acknowledged that forestry should be useful to society but disagreed with the particular solution the coterie advocated. Not all foresters— not even all federal foresters— thought alike, but as one respondent pointed out, "there is no moral tragedy in disagreement among scientists and technical men."[78] Franklin W. Reed, the executive secretary of the Society of American Foresters and also an industrial forester employed by the National Lumber Manufacturers Association, wondered why forestry was even thought to have an obligation to provide service to the social good. He asked, "Is forestry no longer a science, based on a search for the truth?" Reed noted that among the signatories were several "men of independent means," who were "working at some phase of forestry for the pure love of it, [and] . . . absolutely free to express their own ideas and opinions." The wealthy and powerful Pinchot was undoubtedly Reed's main target with this statement, although it also encompassed the well-connected and comfortably retired Ahern as well as Marshall, who a year earlier had inherited the wealth, if not the power, of his famous father, Louis. Among public foresters, many "are free to develop theories of how the forest owner should manage his property . . . for the public benefit, and are not personally concerned, as is the forest owner himself, with the pressing problem of making the property pay dividends." The rest of the seven signatories fell into that category.[79] Academics, too, could afford to work on a social or ecological approach to the forest without concern for their jobs. The freedom they enjoyed differed, Reed argued, from that of foresters working in industrial employ. Foresters who worked for industry "must look upon forestry as a business proposition, to be practiced with a due regard for financial profit, rather than a public cause to be striven for with something akin to religious zeal."[80] While "the forest idealist," such as the signatories of the letter, could afford to "see things as they ought to be" and advocate for public forests, the "forest pragmatist" had no such luxury. Such a person had to "take things as they are" and focus on providing solutions for forest users. While there was room for both idealists and pragmatists within the profession, Reed concluded, it was improper for the idealists to demand that everyone become similarly zealous.

The "Letter to Foresters" discussions illustrate professional foresters' confusion over their debt to science, and over scientists' responsibilities to the public. For the Zon coterie, increasing public regulation, and ultimately public ownership of the nation's forests, was the highest social good. Foresters like Reed, on the other hand, saw social good in helping industry. Creating a better program of industrial forestry would help stabilize forest use, which would indirectly serve the larger public good. Further, as Reed and others pointed out, ecology was not the only kind of scientific research foresters could use in their work. The Zon coterie's arguments about the scientific value of natural, unspoiled forest assumed that proper scientific inquiry would be ecological. However, much industrial forest research could be conducted without any use of ecological methods and theories. Reed made a valid point when he observed that foresters who were academically employed or otherwise "independent" were freer to advocate radical social positions. The problem-solving research of government and industrial foresters demanded more conservative, "pragmatic" conceptions of social utility. Throughout the debates, people used the term "forestry" to describe what they judged good research and practice. However, different foresters used this same term to connote different social goals and research aims. Almost all professionals working in the forest claimed to be "doing forestry" despite vast differences in the practices they referred to with the term. While occasionally a lumber executive would publicly claim to be "doing forestry" as an intentional strategy of obfuscation or misdirection, in most cases it was simply myopia.[81]

The Forester-Activist: Marshall's Sharpening Critique of Forest Use

As the Zon coterie tried to set the course for the future of the profession, Bob Marshall was developing into a powerful voice for the future of the forest itself. While he had already published numerous articles on forests and wilderness, by 1930 his work was drawing more attention. Perhaps his most widely read call for wilderness was an article entitled "The Problem of the Wilderness" in the popular *Scientific Monthly* magazine. "The wilderness is in constant flux," Marshall wrote, and "it exhibits a dynamic beauty."[82] In the article, Marshall argued that wilderness is valuable not just for its aesthetic and recreational worth but also for sheltering the most appropriate sites for ecological study. While in many ways wilderness appreciation approximates art appreciation, the aesthetic value of the wilderness rests not just in trails, vistas, and views but in the dynamic array of animals and plants living there. "A seed germinates, and a stunted seedling battles for decades against the dense shade of the virgin forest. Then some ancient tree blows down and the long-suppressed plant suddenly enters into the full vigor of delayed youth, grows rapidly from sapling to maturity, declines into the conky senility of many centuries, dropping millions of seeds to start a new forest upon the rotting debris of its own ancestors, and eventually topples over to admit the sunlight which ripens another woodland generation."[83] While some had imagined wilderness to be devoid of all human industry, in this essay, Marshall included ecology within the realm of acceptable wilderness activity. The high moral and social value of scientific study, he argued, is on a par with the ability of wilderness to inspire great action, create states of aesthetic bliss, or provide physical challenge and intellectual stimulation. Study of the forest can exist without relating it to economic use of the forest. For Marshall, these ecological interactions were part of the value of wilderness. Such ecological study could be done in harmony with the values of the wilderness and also existed as a personal expression of that value. The article sketched a vision of an ideal state of interaction between forest ecologists and the forest, with no interference from industry.

Although Marshall implicitly condemned the Forest Service for not acting more decisively to protect wild forests, the agency had in fact been working on that issue. The planned Research Reserves would set aside areas of pristine national forest as specimen areas for scientific study. Responding to "The Problem of the Wilderness," chief forester Stuart pointed out that the agency was already in alignment with Marshall on most points he raised. "The Forest Service fully appreciates the social importance of the wilderness, . . . [although] it is a little early as yet to give you specific information regarding the progress made in carrying this policy into execution." He could, however, point to one area of concrete achievement, that of safeguarding and prioritizing scientific research on wild forests. Stuart assured Marshall that "the nucleus of the system of Research Reserves will be established during our current year [1930]."[84] The program of Research Reserves had begun by March 1929, when Stuart had observed that "in late years the value of unmodified areas to science and research, particularly forest research, has become more and more evident and is receiving increasing emphasis from individual research workers."[85] Stuart acknowledged that while much of the national forest land was still in "original or virgin conditions," the threat of logging, expansion, and recreation in the national forests "threatens ultimately to completely modify and transform such areas, reducing them to a common level with the other occupied parts of the United States." This would be an irreparable loss for science, he stated. "Scientific study of the laws of nature will be less effective if the opportunity no longer exists to contrast the operation of such laws within areas subject to human modification and those not so modified."[86] The program of designating and establishing Research Reserves was formally adopted throughout the national forest system by midsummer of that year. In an amendment to the *Forest Service Manual* that went into effect on July 22, 1929, division foresters were instructed to define such areas "to insure the preservation of virgin areas typifying all important forest conditions in the United States." Within these reserves, "scientific and educational use will be dominant. Industrial utilization of resources will be prohibited . . . promiscuous public use will be incompatible with the scientific use of the area." However, Stuart's move somewhat outpaced the law. While the amendment admitted that "present laws are inadequate to permit total exclusion of the public, or prevent the location of mining claims" or easements, "once the required system of Research Reserves has been defined legislation to safeguard them should be obtainable from Congress."[87]

In 1930, while in Alaska, Marshall received a job offer from Earle H. Clapp. Clapp had been recently appointed assistant forester and placed in charge of the newly consolidated research branch of the Forest Service. Clapp thought that Marshall, despite his recent completion of his doctorate, could be extremely successful working in forest policy instead of forest science. Clapp wrote Marshall, "As you know, the question of forest policy is very much alive now. . . . We want to be prepared to help safeguard the public interest and to contribute to it in a constructive, aggressive way."[88] Marshall's recent prominence in the dialogue on public control and his eloquence on wilderness and use in the national forests had brought him to Clapp's mind for a position setting guidelines for federal forestry. Marshall possessed the "capacity for observation and analysis, and also a good deal of imagination," Clapp told him in the job offer. Marshall, now possessing both a doctorate in plant physiology and a national reputation as a forest advocate, was at a career crossroads. Clapp observed to Marshall that he would be uniquely qualified for the position since "you are interested in the field of economics and forest policy and in the social side of forestry, regardless of the fact that your graduate work has been in the biological field."[89] Clapp may have suspected that Marshall envisioned himself as a policy gadfly rather than as a full-time scientist. Certainly, the fervor of Marshall's recent articles had made more of an impact in the public sphere than his scientific writings had. Despite the doctorate, his recent output suggested that Marshall might not harbor goals of devoting his life to research.

As Clapp had surmised, on his return from Alaska Marshall would find employment in forest policy instead of scientific research. While Marshall spent the next two years focused mainly on his Alaska book, he became increasingly involved in forest policy as well. He accepted a position as a Forest Service collaborator, the category reserved for those who held particular expertise that the Forest Service contracted for work on particular projects in a nonpermanent fashion. The collaborator agreements culminated in the publishing, as widely circulated government documents, of the findings of those experts contracted on a particular topic.[90] The design of the position allowed Marshall to remain independent from the daily duties of employment in the Forest Service, giving him the freedom to continue other pursuits, such as seeing *Arctic Village* to publication. Marshall's main duty as a collaborator would be working on the massive Copeland Report. The Copeland Report was a Forest Service effort to compile a comprehensive national study of the nation's forests, industrial forest impacts, and the public control of private forest practices.

An eight-author, 1,677-page document, the Copeland Report was officially titled *A National Plan for American Forestry* and published in March 1933. This study had been ordered in March 1932 by a Senate Committee on Commerce headed by Royal S. Copeland, senator from New York and vice president of the American Forestry Association. The Copeland Report, although requested by the Republican-led Senate during the Hoover administration, nonetheless became the starting point for much of the Roosevelt administration's forest policies. It was an exhaustively multifaceted portrait of the past, present, and future of American forests, covering topics from taxation to recreation to the ecology of forest watersheds. While not the first such congressionally mandated examination of American forests, it was nonetheless the first in many years. Unlike previous such documents, the Copeland Report focused on the administrative structures of the Forest Service and the agenda of forestry rather than on timber supplies and economic conditions. With a great deal of introspection and self-analysis, the report prescribed a program of increased public ownership and regulation by which the government could improve the long-term quality and stability of the nation's forests. The newly appointed secretary of agriculture, Henry A. Wallace, responsible for overseeing many of the Roosevelt administration's New Deal programs, presented the report to Congress. Acknowledging the importance of the report, Wallace prefaced it by stating that "the forest problem [is] one of the largest which the American people have ever faced, and one of the most urgent now demanding attention."[91] Wallace emphasized the importance of implementing the report's prescriptions, writing that "the laissez-faire approach and avowedly planless policy of private ownership is failing to meet the situation. The long-time character of forestry itself, and the impossibility of doing immediately everything which must be done emphasizes the desirability of national planning."[92] The Copeland Report was not intended for government officials; the two-volume hardbound copies of the report circulated widely within professional forestry circles as well. Both detractors and proponents declared the publication of the Copeland Report to be one of the most important developments in American forest policy in years. An editorial in the journal *American Forests* stated that while "some of its conclusions will be questioned and its program of action will be subjected to argument . . . this does not lessen its value. No great public question is ever rightly solved until the question itself is clearly defined and the facts marshalled. This the Copeland report has done in a highly efficient and impressive way."[93] The report's prescription for corrections to the nation's forest problems, however, would be a subject of discussion in forestry journals for months and years to come.

Marshall wrote three articles for the Copeland Report, two on recreation and one on the forests managed through the Department of Interior in the national parks and national monuments. Marshall developed his concept of the interrelations of forest science and forested wilderness more fully in the Copeland Report than anywhere else in his writings. He expanded on the idea he first broached in "The Problem of the Wilderness" that scientific investigations of the forest were on the same level with using wilderness for its scenic beauty or physical challenge. He wrote, "Some visitors to the forest are primarily interested in its scientific aspects. They want to study the forest, to learn the fundamental reasons for its development, to appreciate the causes of the functioning of its myriad component parts. To them the forest is a laboratory, unbounded by the conventional four walls, floor, and ceiling of the usual research center but fully as significant in the development of a knowledge of the laws of nature."[94] He envisioned the wilderness as a place not only for the older tradition of amateur natural history but also for cutting-edge ecological investigations. He also assisted Clapp in editing and compiling the rest of the contributions for the report. Marshall brought to the Copeland Report a perspective on recreation that was mostly missing from the ranks of regular Forest Service employees. Its contributors included a number of members of the coterie, with Zon, Munns, Clapp, Sparhawk, and Marshall each contributing at least one chapter. The clear message of the report was a recommendation for a dramatic increase in public control over forestry practices on both public and private lands, and, as much as possible, a consolidation of that power within the Forest Service. The chapters of the Copeland Report added up to a complete view of the state of the Forest Service and American forests, and a unified set of recommendations for future directions in agency administration and forest management.[95]

The Copeland Report pushed for a more prominent role for foresters and other scientists in public lands policy making. Several chapters explored ways to limit logging and evaluated the noneconomic value of national forests to the nation. Marshall and others wrote about the possibilities for preserving some forests for the nascent Research Reserves program or for other wilderness designations. Research applied to the economics of forestry was often aimed toward saving more of the products of logging from wastage or toward more efficiently using sawmill waste. Such research, to be centralized in the Forest Service's newly built Forest Products Laboratory, would allow production of a greater amount of wood products from the same area of logging than would have occurred before. Products like plywood and chipboard could be created from small trees, scraps, and lesser species and could then often substitute for more basic forms of wood. Likewise, research into specific techniques and procedures for reforestation, both artificial and natural, could establish reliable second growth in areas already logged. Regrowth and subsequent harvests on those areas could alleviate the pressure to harvest remaining intact forest areas.

In many ways, the Copeland Report harked back to Zon's original vision of forest research in the Forest Service. From the beginning Zon had fought for the independence and impartiality of federal forest scientists, both to improve the quality of their science and to limit administrative interference in their function. Research and administration were termed "two distinct classes of work,"[96] with different objectives, practices, and approaches, even if the ultimate goal for both was the maintenance of national forests. Earle Clapp, who led the Copeland Report team, underscored the importance of having researchers free to criticize their agency's management practices if they were scientifically unsound. Forest research, even under Forest Service auspices, should be no different in practice than any other scientific research, he wrote, and "the importance of freedom in research to reach conclusions based on facts cannot be overestimated."[97] The biggest alteration of Zon's original design was to move much of the permanent research staff away from the experiment stations. The initial arrangement had meant "that men might live with their jobs in the woods or on the range . . . it was found, however, that small groups of men stagnated scientifically under such conditions."[98] While the stations themselves remained, moving the research staff closer to cities and universities fostered better circulation of ideas.

Forestry after the Copeland Report

As Marshall wrapped up his assignments for the Copeland Report, and *Arctic Village* went to print, he began to write a new book. *The People's Forests*, completed in 1933, advocated not just increased federal control of forest use but an actual federal takeover of private forestlands. Much of *The People's Forests* was a refinement of ideas on wilderness and recreation Marshall had already developed in his previous essays. He also acknowledged the importance of the Copeland Report as an inspiration, writing that it "not only contains by far the most detailed and accurate statistics ever gathered on American forest conditions, but also advances the most progressive interpretations."[99] Without the restraints of coauthors, Forest Service supervisors, or journal editors, however, Marshall could be more daring in this book than he had been in the Copeland Report or elsewhere. Some of the book echoed "The Problem of the Wilderness" in its celebration of strenuous outdoor recreation and its appreciation of forest aesthetics. One chapter, a "Biological Interlude" describing a forest's ecological interactions, made the case for a scientifically informed perspective on the forest. However, the bulk of the book concerned his recommendations that the federal government turn away from public regulation in favor of dramatically increased public ownership of forests. He had begun this line of thought in his little-read 1930 pamphlet, *The Social Management of American Forests*.[100] While that essay had advocated socializing forest management, it stopped short of fully pressing for public ownership. In *The People's Forests*, however, Marshall proudly declared his radical left allegiance, stepping closer to true socialism. One must note, however, that he advocated purchase of private forests, not nationalization. Marshall consistently underscored that he was in favor of a gradual socialization, in contrast to the recent violent successes of revolutionary socialism overseas.[101]

Although *Arctic Village* sold much better, the publication of *The People's Forests* was a more important accomplishment to Marshall. He even remarked to Zon that *Arctic Village* was, to him, nothing more than an advertisement, a way for him to raise public recognition of his name and personality. Once he had people's attention, he could direct them toward his real work, *The People's Forests*. Marshall dedicated *The People's Forests* "To George P. Ahern, Earle H. Clapp, Edward N. Munns, Gifford Pinchot, Raphael Zon: Courageous foresters who for years have been battling effectively and uncompromisingly for the social management of American forests." Aside from Clapp, all of these men had been signatories on the "Letter to Foresters" three years earlier. For Marshall, this was the inner circle of the coterie pressing for change both within the profession and in the world of policy. His inscription both reinforced the bonds that existed among them and reinvigorated the sense that there was a small but elite group that stood in opposition to the agenda of mainstream forestry.[102]

Conclusion

This chapter has examined the development of ecological approaches in forestry. Since ecology focuses on untouched, nonindustrial environmental settings, ecologists are inclined to advocate for the preservation of the wild state of the places they study. The chapter followed Bob Marshall as he developed from young forester into ecologist, and from ecologist into adventurer, activist, and policymaker. Marshall joined a group of elder statesmen of forestry to create an intellectual coterie that was disaffectedly critical of the direction of forestry and enthusiastically activist about the possibilities for the future. This group of foresters set the tone for the wilderness activism and professional criticism that developed during the New Deal years. The Copeland Report was a document by and for the Forest Service, but the reformist mentality it represented expanded to other federal land management agencies as well. Marshall's writings had critiqued the Department of the Interior's national parks and monuments forest management as well as the Department of Agriculture's national forests. Indeed, the Department of the Interior was widely disparaged for mismanaging the public-domain forestlands entrusted to it, which included the Indian reservations and the General Land Office as well as the parks and monuments. The secretary of the interior, Harold Ickes, resolved to turn a new leaf for the Interior and, as part of this statement, hired Marshall. Marshall, now well known as a radical critic of standard federal forestry, became the new forester for the Office of Indian Affairs. With that bureau, Marshall would administer forest policy for each of the reservations, working with often underprivileged tribes to create beneficial arrangements of forest use and preservation. For Marshall, this was a unique opportunity to serve the social good as well as a way to avoid tangling with the demands of the logging industry. The new job was a departure from the Forest Service, and it also meant he had taken a permanent departure from the realm of forest research. Instead, he traveled around the nation, an administrator and policy maker for the vast variety of forestlands scattered from coast to coast within the reservation system.[103] Despite his departure from the Forest Service and from scientific research, he would continue to be a thorn in the side of the Society of American Foresters. He continued to fuel disputes with contributions sharply critical of both the Forest Service and the status of the profession of forestry.

Upon its publication, the Copeland Report had seemed to many to be the first battle in what they envisioned would be a revolution of forestry. The year of its publication, 1933, was a high-water mark of the New Deal, a year that marked a sea change in federal management on all levels. Against a backdrop of changes in federal policy, the early 1930s also provided a welcoming atmosphere for the articulation of radical critiques of professional forestry. This had an immediate and widespread impact on the community of foresters, creating intense discussion of the importance of both federal control and research within the Forest Service. In 1934, Aldo Leopold looked back on the changes in professional forestry during the previous year. He compared the looming weight of the Copeland Report to the creation of the moon itself from a fragment of Earth pitched skyward by the gravitational pull of a passing celestial body. "Conservation, I think, was 'born' in somewhat the same manner in the year A.D. 1933. A mighty force, consisting of the pent-up desires and frustrated dreams of two generations of conservationists, passed near the national money-bags whilst opened wide for post-depression relief."[104] As the next chapter will show, the struggles between the conservationist impulses of leftist foresters and the pressures for industrial development of forest resources were far from over.

4 Forestry, Wilderness, and the New Deal

The Tillamook State Forest lies in the Coast Range of Oregon, a short drive west of Portland. Generations of Oregon children, in school groups and scout troops, have made excursions into this forest for nature study and recreation. Crisscrossed with hiking and off-road-vehicle trails, in easy reach of urban denizens searching for a place to play, the Tillamook has served as one of the most accessible examples of the state's great forest wealth. Lacking the stunning scenery and rugged high country of the Cascades forests, the Tillamook has value for many Oregonians today for its rolling landscape and gentle climate. At the end of the 1920s, this landscape was treasured for its aesthetic beauty, but for other reasons as well: it held some of the most valuable timber in the state. The Tillamook's dense stands of Douglas fir attracted significant lumber activity, and by 1930 loggers were delving deeper into the interior of the Tillamook, as narrow-gauge rail allowed them to haul their steam-powered donkey engines and other logging equipment deeper into the forest. Still, the Tillamook was rugged and difficult to traverse under any means. No navigable rivers meant moving overland through dense undergrowth and over steep terrain. Many who arrived in the area during the early years of the Depression, hoping for logging work as other jobs disappeared, found their days full of backbreaking labor. The difficulty of bringing logging equipment into the stands of timber was more than matched by the struggle of getting the timber out of the hills to mill and market.

On August 14, 1933, a dry and windy day, the Oregon Department of Forestry sent word out to logging operators in the Tillamook to suspend operations for fear of a blaze. Some of the small "gyppo" crews, trying to eke out a living logging the salvage from better-established lumber operations, ignored this warning and continued to work. The effects of the Depression had made the already-unstable gyppo logging operations even more financially precarious, and these loggers were loath to lose even a day's worth of work. A spark from a steam donkey, skidding a log through slash, started a fire in the kindling-like waste. On the hot and dry hillside, this blaze leapt out of control almost immediately. By the end of the day this first blaze was joined by several others scattered through the thick forests of the Tillamook.[1] With dry weather forecasts meaning there was little hope of rains extinguishing the fires, the managers in the Oregon Department of Forestry, who controlled the Tillamook, marshaled what forces they had to fight

Gifford Pinchot late in 1922, smiling during the lull between his successful gubernatorial campaign and his inauguration as Governor of Pennsylvania. While his two non-consecutive terms as Governor (1923-1927 and 1931-1935) broadened his focus, he maintained a lifelong interest in the forest industry, federal forest policy, and professional forestry. Library of Congress, LC-DIG-NPCC-07503.

the danger. They quickly realized the fires were too complex and extensive for them to handle alone, and they turned to other government entities for help. A nearby camp of forest workers from the Civilian Conservation Corps, which had only recently been established, provided additional labor to the firefighting force upon request from the state of Oregon.

The fires finally abated on September 5, as a soaking rain in the end accomplished what the concerted effort of hundreds had not been able to do. The fires had left behind a deeply charred landscape, silt-choked streams, and a thick layer of ash on surrounding farmland. By the time all crown and ground fires came under control eleven days after they had started, they had damaged more than 235,000 acres of high-quality Douglas fir forestland, much of which had still never been touched by logging activity.[2] As devastating as that was, the aftermath of the fires would become more severe and long lasting than anyone anticipated. Industrial landowners abandoned their now worthless lands, defaulted on their taxes, and let the forests revert to state ownership. Soon well over half of the Tillamook forest was publicly owned but languished untended as new growth haphazardly filled in the charred areas. Local communities, economically devastated by the fires, began a process of salvage logging to extract any remaining merchantable timber from the scorched lands. The 1933 fires were devastating to the local economy and the state at large, but the effects of the initial fire year were exacerbated by severe fires that reburned the newly growing forest in 1939 and 1945. With each successive fire, an even greater proportion of the vegetation that grew in consisted of the brushy undergrowth that choked out conifers. Twelve years after the initial spark caught flame, the Tillamook was as far from its initial glory as it had ever been.

Public outcry following the 1945 fires spurred Oregon politicians to fund a special committee that recommended intensive intervention in the area to ascertain that a more permanent forest would become established. Beginning in the 1940s, the state funded a massive reforestation program, which was carried out to a great extent by volunteer labor.[3] Portland high schools, which a generation earlier had sent their students to the Tillamook for recreation, now sent them there to plant seedlings.

Today's Tillamook bears the legacy of these fires, their salvage, and the intense interventions by state government to reestablish the forest afterward. The circumstances of the initial 1933 fire and the confusion over how to administer a functional program of rehabilitation following its devastation show the unsettled state of forest policy and understanding during the New Deal. Firefighting, fire prevention, and reforestation required coordination between the well-established federal and state agencies and the project-oriented ones of the New Deal. The history of the Tillamook demonstrates the struggle over defining public forest policy, and the growing role of the economics of lumber in setting the agenda for policy in the region. In the multiple reassessments of the predictions for regrowth we see a new acknowledgment of the difficulty of fully understanding the patterns of regrowth of forest species following disturbance. Northwest residents felt cultural attachment to their forests not only for employment and economics but for recreation, aesthetics, and regional identity. The Tillamook shows the solidification of regional identity around the Douglas fir in particular, as a cultural symbol as well as an economic engine. The Tillamook fires inspired the citizens of Oregon to cement their relationship to the forests that surrounded and sustained these communities.

The history of the Tillamook reflects the multiple forces acting on Northwest forests as they became increasingly important, both economically and culturally, during the 1930s. Given that the New Deal increased the power and reach of the federal government's land management, one might assume federal foresters welcomed its changes. However, the restructuring efforts of the New Deal were protested from within the ranks of foresters just as they were by those in industry or in state government. Foresters faulted the New Deal for adding new layers of complexity to already complex land management bureaucracy, for siphoning money and personnel away from preexisting agencies, and for creating confusion over goals and agendas. In the end, the programs and policies of the New Deal moved the federal forestry agenda significantly in a direction that many foresters themselves did not want: away from general science and toward problem solving and profit.

This chapter discusses the changes within the profession of forestry during the New Deal–era upheaval of federal agencies. The New Deal challenged foresters to reconceptualize their profession as both the lumber industry and federal public lands management agencies changed dramatically. The previous chapter examined the formation of a coterie of radical foresters, headed by Bob Marshall, Raphael Zon, and Gifford Pinchot, who wanted to make their profession both more ecological and less industrial in its orientation. While adopting an ecological approach could enrich and sustain forestry's scientific value, it also had the potential to steer the profession farther away from its roots in social utility. The current chapter describes how the members of the Zon coterie intensified their criticism of their profession. In June 1934, they drafted a document that came to be known as the Zon Petition, which demanded significant change in the goals and practices of the Society of American Foresters. Energized by Washington's pervading atmosphere of change, the coterie urged their colleagues toward an approach that was both socially informed and scientifically sophisticated. Rather than filling its pages with small-scale research focused on matters of industrial yield and production, they demanded the *Journal* cover basic research, social forestry, and forest management. However, increasing numbers of foresters were employed, directly or indirectly, in helping the lumber industry function. The radical foresters sought a different way to marry their profession's interests in utility and research. The new conception of forest wilderness as an aesthetic and recreational category came to offer another avenue for these disaffected professionals. Land-use activism, as embodied in the new Wilderness Society, shunned industry connections but still focused on management and utility. Wilderness activism became a way to envision a future for forestry and forest science that did not require capitulating to industry needs or government guidelines. This chapter explores how disagreements about professional priorities within forestry in the mid-1930s were the spark that lit a wilderness movement that would catch fire in the decades to come.

The New Deal and Its Discontents

Franklin D. Roosevelt's New Deal, programs of social and economic transformation, grew quickly in the months after his inauguration in 1933. The New Deal touched almost every aspect of American culture and society, and forests were no exception. The scope of the New Deal's alterations of federal and state forest and land-use policies was massive, and its changes deeply affected the practice and culture of forestry throughout the country. Not only did the

Roosevelt administration overhaul the existing agencies and programs concerned with forests and forestry, it also instated a number of new programs and agencies to tackle various forest issues.[4] The Roosevelt administration aimed to rescue the nation from the economic depression by whatever means were necessary. Roosevelt was less concerned with political orthodoxy than with providing workable solutions to acute problems. The programs he advocated to combat the nation's ills were often experimental, relying on untested methods and strategies. His programs were not centered on a particular economic theory or model for recovery so much as on a willingness to adopt any idea that would lead to success. If he had any real economic tenet, it was that governmental debt was to be avoided at all costs and used only when all other options had failed. However, this economic conservatism did not keep Roosevelt from launching a wide variety of programs designed to help practically every facet of the American economy. He was willing to invest great amounts of federal capital in programs such as the Shelterbelt Project, for which precedents were few and predictors of success unreliable. Such tactics meant that the New Deal inspired much criticism among both Roosevelt's political adversaries and the general American public.

The sheer variety and extent of these New Deal forestry programs implied that deep change was afoot for many traditional practices in forest administration and organization. In addition, foresters foresaw the projects and employment patterns of their discipline changing as they responded to the new demands placed on their profession. Members of the Zon coterie were involved with the New Deal from the beginning. Pinchot had been tapped to design new federal forest use policies very early in Franklin Roosevelt's administration. Pinchot accepted this assignment, although as the governor of Pennsylvania, he had little free time to devote to it. Also, despite his involvement with the Zon coterie, Pinchot was not fully versed with recent advances in conservation theory and forest management policy. He called on Marshall, his friend and a man of similar beliefs, to help prepare the policy. The plan Marshall presented to Pinchot, which Pinchot then presented to Roosevelt, was very similar to that laid out in *The People's Forests*. Pinchot and Marshall's main conclusion was that "private forestry in America, as a solution of the problem [of forestland devastation], is no longer even a hope. Neither the crutch of subsidy nor the whip of regulation can restore it. The solution of the private forest problem lies chiefly in large scale public acquisition of forest lands."[5] Like *The People's Forests*, Pinchot and Marshall's plan owed a large debt to the ideas of Raphael Zon. While preparing the documents for Pinchot to show to Roosevelt, Marshall wrote to Zon that "lots has happened in connection with the Forest policy during the past week though I do not know for sure whether that lots will amount to anything, or meet with your approval." Marshall's description

shows how influential the coterie's opinions were in the construction of New Deal
forest policy:

> In re-submitting it [the forest plan] to Roosevelt, he [Pinchot] made
> numerous changes in language and paragraphing but did not alter a
> single one of the ideas or add any new ones. He did have to insert one
> or two extra "devastates," insisted upon starting things off with the
> 822 million acres of original forest and made me rewrite the second
> regulation in briefer and stronger form. It said in fact that under present
> conditions all lumber companies were too near bankruptcy to be forced
> into Forestry through regulation and if they ever became prosperous
> again they would become sufficiently rich to buy off any regulating
> organization that might be set up. When we finally incorporated this,
> Pinchot said with a sigh: "I have argued for Forestry for thirty years
> but I am convinced that you are right—we need something more."[6]

The deeply felt conservation ideals of this cadre of like-minded foresters formed
the basis for the land-use policy of the Roosevelt New Deal, albeit eventually
somewhat blunted by political realities.

A number of New Deal programs directly addressed problems of forests
and other aspects of American public lands and natural resources. Several New
Deal programs mandated changes in the practices and policies of foresters
and backed up these mandates with financial assistance. The administration
of Herbert Hoover continued the commerce-oriented policies in the nation
begun by Harding and Coolidge. In the name of giving more power to the state
governments, individual citizens, and corporations, Hoover was reluctant to
create any new federal bureaucracies or even to expand any federal programs.
Faced with the intense economic crisis following the stock market crash of
1929, the Hoover administration looked for market-driven solutions rather than
administer relief aid through the channels of the federal government. The New
Deal programs recognized that conservation and land-use problems were not
being handled effectively by the individual states. Much of the decision-making
power that Hoover had ceded to the states was reappropriated by Roosevelt
through nationwide programs that supplanted or abolished state programs.
This federalization of forestry allowed foresters formerly working on a state-
by-state basis to move to a nationwide view of the problems and opportunities
of forestry. Effectively, this interest of the federal government in forest policy
brought professional foresters right into the politics of the day. What might once
have seemed simply professional judgments now seemed more like critiques of
government policy. What might once have been mere academic discussions now

became part of the government's new land-use programs. The lumber industry was in flux, the profession of forestry was in flux, and the result was a year of both intense activity and intense introspection among forest professionals. Foresters' recommendations could hold weight in planning decisions in ways they never had before. The feasibility of New Deal forest and land-use plans often hinged on the predictions and projections of foresters.

The New Deal proposal that most directly addressed the conduct of professional foresters and the lumber industry was contained in the provisions of Roosevelt's National Recovery Administration, or NRA. Roosevelt created the NRA in conjunction with the Roosevelt-backed National Industrial Recovery Act (NIRA) in May 1933. NIRA was intended to financially stabilize the nation's major industries through regulation of nearly every aspect of their functioning. Many industries were still in a shambles from the aftermath of the stock market crash of 1929 and were suffering from years of mismanagement and poor planning. The NRA developed both a set of general regulations with which to stabilize American industry in general and a set of codes of fair practice tailored for each of more than five hundred individual industries. Most of the provisions of the NRA addressed two problems industrialists were facing at the time: coming to agreements with labor unions in setting working standards in the industry, and conversing with other managers in the same industry in order to stabilize it without violating antitrust laws. In the beginning, industrialists thus welcomed the passage of NIRA because they believed the resultant agency would greatly benefit their ability to conduct business. Adoption of the individual codes of fair competition was enforced by law, although the creation of the code did allow for some input by a particular industry's leaders. One of the many industries the NRA targeted was the lumber industry, which had been cycling through unstable boom-and-bust cycles for decades and had been in collapse in most timber-producing regions of the nation during the preceding decade. From the inception of the NRA, many in industry chafed under the codes that were imposed on them and questioned the NRA's constitutionality. However, even those who believed that the NRA was unconstitutional and hence only provisionally important were forced to obey its strictures while they held.[7]

The Lumber Code of the NRA was designed with a great deal of input from the lumber company owners of the Pacific Northwest and other regions of the country. The Lumber Code went beyond the labor and competition concerns to which most of the other industry codes restricted themselves. Article X of the Lumber Code was of particular interest to foresters, as it concerned resource conservation through industrial self-regulation. Many in the lumber industry realized that without vocal protests on their part, the code for each of their

Raphael Zon affected national forest policy even when his office was far from Washington, D.C. Zon, photographed here circa 1935, directed the Lake States Experiment Station in Minnesota from 1923 until his retirement in 1944. Courtesy Minnesota Historical Society.

industries might become a hindrance to business as usual. As Mark Reed, the president of the newly formed Pacific Northwest Loggers Association, put it, "unless we govern ourselves, we are going to be governed, and, in all likelihood, in a way we do not like."[8] The drafting of the Lumber Code was attended by logging and lumber milling industry representatives from around the country, including the Pacific Northwest. The lumbermen intended to head off the gathering forces of Americans demanding preservation of forestlands. F. E. Weyerhaeuser voiced the concern of many lumbermen during the summer of 1933 when he stated that "undoubtedly there will be a tremendous drive made by our emotional forester friends who think there is rare opportunity to force upon the lumber industry all of their theories and imaginings" about forest preservation.[9] The *Journal of Forestry* featured an in-depth series of articles about the Lumber Code of the NRA. Beginning with the passage of the NRA in 1932 during the last months of Emanuel Fritz's editorship, the *Journal* chronicled the Lumber Code through every stage of its creation and implementation, eventually documenting its demise in May 1935. These articles favored self-governance of lumber industry conservation efforts and downplayed the need for federal regulation. Some read these as biased against the Roosevelt administration, a sentiment amplified by the known affiliation of their author with the lumber industry. The chief forester for the Long-Bell Lumber Company of Longview, Washington, John B. Woods, also wrote several articles on Article X for the *Journal*. In his role of secretary of the Joint Committee of the National Article X Conference, Woods was charged with advising the US secretary of agriculture on the writing and implementation of the so-called Conservation Article of the Lumber Code. While the committee included delegates from the academic and governmental forestry communities, it was dominated by industry-affiliated foresters. Woods's articles were essentially an attempt to gain public support for a committee whose objectivity and strength were already under question by many American foresters.[10]

Perhaps the most widely remembered of these New Deal programs was the Civilian Conservation Corps, commonly known as the CCC. The CCC served two equally important purposes in Depression-era American society. First, it allowed vast numbers of unemployed youth and young adults to find gainful employment rather than live on charity. And second, it addressed the increasing devastation and neglect of the American countryside by supplying a workforce for conservation. Begun in 1933 as the Emergency Conservation Corps, the program quickly became one of the largest sources of labor ever applied toward conservation work. At its height, in October 1935, the corps employed nearly 560,000 men, and over the nine years it was in operation 2.5 million individuals worked in its employ throughout the nation. To distinguish it from the Army Corps of Engineers, the Emergency Conservation Corps was commonly known as the Civilian Conservation Corps, although it was not officially renamed until 1937. The corps was a joint project between the federal Department of Labor, which was responsible for recruiting and administration, and the Departments of Agriculture and the Interior, which along with the Army Corps of Engineers provided the expertise and management of the corps' projects. The corps was notable for the level of cooperation it brought about between these sometimes hostile federal departments, and also for its collaboration with the state agencies of parks and forests. The foresters and others who steered the corps focused on massive tree-planting projects on cutover lands throughout the nation, as well as on trail building, erosion control, and other large-scale projects. At the ceremonies marking the dissolution of the corps, officials estimated that 2.25 billion trees had been planted in various reforestation and afforestation projects, and that improvement projects had affected another twenty-five million acres of publicly owned standing forest. The impact of the Civilian Conservation Corps can still be seen on public lands throughout the nation, in the form of trails, park buildings, and now-mature tree plantations.[11]

Agricultural problems were taken up as a pressing concern during the New Deal.[12] The issue of soil erosion, in particular, affected land management policy decisions in the forests and lumber industry. The federal government first addressed soil erosion under the aegis of the Department of Agriculture's Bureau of Soils,

George P. Ahern lived in Washington, DC, during the latter years of his career as well as in his retirement. For a time he served as secretary of the Army War College Washington Barracks (1922). Photograph 83908; Records of the Office of the Chief Signal Officer, RG 111; National Archives at College Park.

in the form of erosion experiment stations in the farm states of the nation. In the early years of the New Deal, the Departments of Agriculture and the Interior were locked in a struggle for power over lands and responsibilities. As well as a disagreement over policies, the struggle was a battle between the two strong personalities heading the departments, Henry Wallace in Agriculture and Harold Ickes in Interior. Soil erosion prevention became one of the federal programs that were torn between these two departments. Despite a long history of soil erosion projects in the Department of Agriculture, Ickes set up a new Soil Erosion Service in the Department of the Interior and lured Hugh Hammond Bennett, the strong and dedicated head of erosion projects in the Department of Agriculture, to run it. Despite this infighting Bennett made sure the service stayed effective throughout the transition. One of the most important projects Bennett spearheaded, at least from the point of view of foresters, was the implementation of tree-planting projects to prevent soil loss. In most of the soil types of the United States, the weblike root systems of trees and other plants held the rich topsoil in place against the erosive effects of torrential rains and winds. If these root systems were removed through excessive logging or plowing, severe erosion and drying of soils could result. The topsoil, once washed downstream or lost in the wind, could not be replaced or reproduced. Foresters had remarked for a number of years on the importance of vegetational cover to soil quality.[13] Before the Soil Erosion Service, however, no sustained efforts to remedy the problem had been undertaken. With the new reordering of the federal agencies there was an increased opportunity for interagency projects and cooperation.[14]

The Prairie States Forestry Project, better known as the Shelterbelt Project, is a prime example of such cooperation in the name of erosion prevention. In an attempt to save the soils of the Great Plains from being scoured away by winds, a massive, federally funded project was begun in July 1934. Franklin Roosevelt himself conceived this project in the early days of his presidency as a possible solution to the crisis of the Dust Bowl. Roosevelt imagined a belt one hundred miles wide, reaching from Canada south through the Dakotas, Nebraska, Kansas, and the Oklahoma Panhandle and ending in northern Texas, which would contain within it windbreak strips of quick-growing trees aligned perpendicularly to the prevailing winds. The feasibility and efficacy of the project were debated from the beginning, from both a scientific and political perspective. No one had proven that trees in such a planting project would, first, survive the harsh climates, and second, obtain the desired soil stabilization. Also, creating such a shelterbelt would be a tremendously expensive proposition, especially when its value was so dubious. Roosevelt funded it through executive orders, classifying it as part of the national drought-control effort and requesting chief forester Robert Y. Stuart to produce a workable shelterbelt plan. Stuart tapped Raphael Zon to author the Forest Service's plan, and in 1934 Zon officially assumed a supervisory role for the Shelterbelt Project. Zon designed the windbreak plantations from his office in Minnesota while maintaining his other duties. Crews of tree planters, mostly from the CCC, used seedlings supplied by the US Forest Service to establish windbreaks on private farmland free of charge to the landowners. The project required cooperation between individual landowners and states as well as joint efforts between various agencies of the Departments of Agriculture and the Interior. While this interagency cooperation may have benefited the Shelterbelt Project in some ways, it was not the project of any single department and therefore its funding was always in jeopardy. The total number of individual farmers who participated in the project is estimated at thirty-three thousand, all of whom agreed to maintain the shelterbelt at their own expense after its planting. The program eventually fell victim to its funding woes and was discontinued in 1942, its effectiveness in erosion control still in doubt.[15]

The Zon Petition and the Fracturing of Professional Forestry

The New Deal created a cultural milieu of governmental and social change. The Cabinet expanded, departments changed their missions, new agencies were proposed and created, and officials reassessed existing divisions. This mood of reorganization and change affected all parts of the executive branch. Scattering new conservation agencies throughout multiple departments created competition and overlap, but that was not necessarily bad. As Bob Marshall cynically observed about the proposed creation of an overall Department of Conservation, "a large faction of cabinet officers are bound to be either stupid or crooked. However splendid certain cabinet officers may be at a given moment, I think in the long run it is much safer to have the 'growing crop conservation activities' of the government split between competing, jealous departments, as long as we have huge concentrations of wealth to encourage the government to be dishonest."[16] Turmoil and mutual distrust fostered necessary conscientiousness. The state of federal agencies in these years was reflected in the confusion among American foresters, whose professional identity was closely tied to governmental conservation and resource management. The question of how closely forestry was tethered to social utility became more prominent as foresters became more important for both private and public land management decisions. The programs of the New Deal were intended not only to lift the country out of its economic depression but to fix the problems in the nation's infrastructure through experimentation and innovation. In this atmosphere of enlightened improvisation, scientists found themselves in positions of political power as their expertise became the basis for policy innovations. In the mid-1930s the Zon coterie was motivated and inspired by the spirit of scientifically informed activism that drove New Deal innovations.

In the summer of 1934, an unprecedented document was sent to the members of the editorial council of the Society of American Foresters. Written in the form of a petition, the document appealed for changes in the editorial policies and practices of the *Journal of Forestry*, the official organ of the society and the foremost American journal in the field of scientific forestry. Conceived by Bob Marshall and Raphael Zon, the petition was, in some ways, a follow-up to the "Letter to Foresters" the Zon coterie had produced in 1930. While the "Letter" had critiqued the general shape of the profession, this petition, on the other hand, targeted the *Journal*, and by extension the leadership of the Society of American Foresters. Signed by twelve of the most prominent professional foresters in the nation, this petition demanded large-scale changes in the content and function of the *Journal*.[17] What came to be known as the Zon Petition, after its last signatory, Raphael Zon, was a watershed moment in the history of American forestry, both a sign of its state of crisis and a spur toward further self-evaluation. The Zon Petition was a symptom of the increasing split between those who focused on ecological holism and the protection of remaining untouched wilderness, and those who focused on improving methods for using American forests. The intensity of the debates it engendered shows how seriously foresters from all sides of the spectrum took these issues. In essence, this was a fight between those who believed forestry's goal should be better solutions for the technical and management problems raised by logging, and those who believed that forestry's goal was to maintain and protect wild forests for their own sake.

The signers of the petition were some of the best-known foresters in the country, as well as some of the most vocal proponents of conservation and public land ownership. All twelve of the petitioners were influential figures in the Society of American Foresters and in the field of forestry at large. Nine of the twelve held high-level federal government posts at the time of the petition, making the line between personal opinion and governmental policy difficult to discern in the petition. The petitioners included the central Zon coterie, of course, consisting of Raphael Zon, Bob Marshall, Gifford Pinchot, and George P. Ahern. The rest of the signatories were mostly career Forest Service employees. Ferdinand A. Silcox, who had assumed office as the chief of the Forest Service in 1933, was a signatory. Earle H. Clapp, at the time the head of the Forest Service's research division, signed, as did Carlos G. Bates, Zon's research collaborator and coauthor of the book *Research Methods in the Study of Forest Environment*.[18] Also appearing on the list was forest economist William N. Sparhawk, another Zon collaborator and coauthor of two publications about worldwide forest resources.[19] Leon Kneipp, formerly with the National Conference on Outdoor Recreation, was working in the Forest Service's land acquisition program at the time of the petition. W. C. Lowdermilk, director of the Bureau of Soil Erosion Control in the US Department of the Interior, was the only other signatory besides Marshall employed at that time by a federal agency other than the Forest Service. The last two signers, Edward N. Munns and Edward C. M. Richards, were also career Forest Service employees. Ward Shepard was the only person who had signed the 1930 "Letter to Foresters" who did not reappear here. In 1933 Shepard had moved into a position as an American liaison for the German-backed Carl Schurz Memorial Foundation. Shepard's move had precipitated a private, but severe, falling-out with Zon, Marshall, and other friends upset by the increasingly repressive and xenophobic political climate in Germany.[20]

Raphael Zon had become disheartened by the profession's preoccupations with technical planting problems at a time when so much change was afoot. Zon remarked to Pinchot that "'Stop Forest Devastation' is still the banner under which we foresters must march. I cannot grow enthusiastic over planting cutover lands when all around us the forest is being destroyed at an accelerating rate."[21] The *Journal of Forestry*, however, reflected the state of the profession by publishing articles about private forest management techniques, some of which would undoubtedly guide those logging the nation's remaining wild forests. While some critical articles also appeared, the studied neutrality of the editorial decisions led the Zon coterie to view the *Journal* as implicit in industry abuses. In essence, the coterie saw the *Journal* as guiding the field, not just reflecting it. By changing the *Journal*, they hoped to change the entire profession. The petition opened with a statement that they had "come to the conclusion that the editorial policy of the *Journal* during the last few years no longer represents the broad social ideals of the founders of the Society." While acknowledging that the *Journal* was "satisfactory" in its coverage of technical and scientific advances in forestry, the petition stated that its coverage of social and political elements of the field was lacking. The petitioners wrote that "at a time when vast, surging forces are overwhelming the accepted truisms of the past, the *Journal* is lost in petty quibbling over inconsequential matters and artificially created issues. The vital problems of forestry are overlooked or discussed not from a social standpoint or in the spirit of the New Deal."[22] They threatened to start a new journal to counteract what they described as the "purposeless editorial policy" of the *Journal of Forestry*. Funding from unnamed sources was already available for such a publication, and such a journal could have split the society and drastically altered the landscape of American professional forestry. Before they took this step, the petitioners appealed to the Executive Council to "view the situation with us dispassionately and realistically, in an effort to make the *Journal* the real spiritual spokesman of the profession."[23]

Perhaps more than most professional journals, the *Journal of Forestry* existed not just to provide a platform for scientific papers but also to define the ethical, political, and social stances of the profession. The members of the Society of American Foresters were a diverse group, including wood chemists and engineers, forest economists and tax experts, forest ecologists, tree physiologists, and park administrators. Many of these professionals had little interaction with the work of their colleagues from other areas of this diverse profession. The *Journal of Forestry* tried to unite the discipline through weighing in on political and philosophical questions confronting the membership of the Society of American Foresters. An issue of the *Journal* typically contained an editorial, several pieces examining or commenting on aspects of forest legislation and politics, several technical papers on aspects of scientific forestry, a section of shorter pieces of either scientific or political content, book reviews, letters to the editor, and a "Society Affairs" section reporting on meetings and membership. While the *Journal* was the organ for disseminating scientific information on aspects of forestry, it also observed and commented on the forest, land, and industrial policies of the government, and on governmental relations with scientific foresters. Within the *Journal*'s pages, foresters engaged in frequent introspection about the nature of their profession and its responsibilities and role in society.

The production of some form of journal had been one function of the Society of American Foresters since its founding in November 1900. In describing the society's stated mission the founders wrote that "the objects of this Society shall be to represent, advance, and protect the interests and standards of the profession of forestry, to provide a medium for the exchange of professional thought, and to promote the science, practice, and standards of forestry in America."[24] The purpose of the society's journal was thus not only to disseminate information on scientific forestry but also to address issues of the social and industrial ramifications of forestry in practice, and of governmental regulations of forestlands and practices. The *Journal of Forestry* also served as the voice of the society, publishing news of the profession and information about conventions.[25] The editorials were usually written by the president of the society or by the editor of the *Journal*, although occasionally others in the society wrote them. Beginning with the resignation of editor in chief Emanuel Fritz in 1933, the editorials were signed by their authors in order to indicate that the opinions of the writer should not necessarily be taken as the opinion of the society as a whole. This allowed for perhaps more diversity of opinion to be expressed in the editorials and opened up the dialogue of ideas in the journal as a whole. For example, an article written by the society's president, H. H. Chapman, two months before the arrival of the petition began with the question "What is the Society of American Foresters?" and went on to examine the role of scientific foresters and the meaning of professionalization in the complex landscape of American forest and lumber industries.[26] In another example, the *Journal* printed an anonymously written lament of the difficulties faced by the new generation of academically trained foresters in establishing their careers.[27]

Under Raphael Zon's editorship, the *Journal* had maintained a focus on public forest issues and scientific study. However, in the years since he had stepped down from the post, both the *Journal* and the profession as a whole had grown closer to and more supportive of the lumber industry and its needs and goals. Emanuel Fritz, a professor of forestry at the University of California at Berkeley, took over as editor in chief of the *Journal* in 1929 after serving as an associate editor for the previous eight years. At the annual meeting of the Society of American Foresters in December 1932, Fritz tendered his resignation, stating that the demands on his time had become too much to bear. There were, however, signs of dissatisfaction and conflict in the society and the *Journal* prior to the petition's arrival. Fritz touched on the contentiousness of the *Journal of Forestry*'s content in his letter of resignation, published in the February 1933 issue. Fritz, referring to himself in the third person, wrote that

> some members of the Society seem to think that the *Journal* is deteriorating when it does not print harangues on the need for forestry or diatribes against lumbermen. The Editor has held to the belief that it is *settled* among the readers that forestry *is* desirable and necessary, and that nothing can be gained by continually arousing the animosities of private [land]owners and that space had best be devoted to actually advancing the technique of forestry practice and the improvement for the possibility of forestry of any kind on private lands.[28]

The disputes at the *Journal* had become more heated as the profession had grown closer to the lumber industry. Increasingly, the disputes centered on the scope and reach of professional forestry, the meaning of the profession in a changing world. Since Fritz's resignation took effect immediately, the society's Executive Council deemed that the executive secretary, Franklin Reed, should act as editor of the *Journal* until the following May, while a search for a permanent replacement took place. Reed, a consulting forester based in Washington, DC, had close ties to the lumber industry. He agreed to serve as acting editor along with continuing his duties as the executive secretary in charge of membership and meetings. The normal difficulties associated with travel and scheduling made it impossible for the council as a whole to meet to discuss finding a new editor, and by that May they still had not found any person both suitable and willing to take the job. In this bind, they asked Reed to continue as acting editor while they searched for a permanent replacement.[29] Fritz's resignation had come at the beginning of the flurry of New Deal activity in the field of forestry, when the constant changes and continuing debates brought forestry issues to the forefront of society's thought in a way they never had been before in the United States. Not only could the committee members not find a replacement with the time and energy for the editorship, they could not even find time to meet among themselves and discuss possible replacements. Despite his ad hoc appointment, Reed took his duties seriously, writing editorials and articles for the *Journal* as well as soliciting articles from others on topics that he felt were underrepresented. In particular, he urged Marshall and other foresters working within the New Deal government to write papers for the *Journal* explaining the methods and goals of their various projects. While he complained later that these articles had not been as plentiful as he would have liked, a number of articles did appear.[30]

The meat of the petition was three suggestions for changes in the operation of the *Journal of Forestry*. The first suggestion was that the offices of executive secretary and editor in chief be separated. As discussed above, this had been a temporary measure for the *Journal*, and one that the Executive Council had been trying to correct for more than a year. The second suggestion was a description of the petitioners' ideal candidate for editor in chief: "a man of high standing in the profession and of scholarly attainments and literary ability—a man with strong social convictions but tolerant of the opinions of others, scrupulously honest intellectually, and a strong believer in freedom of expression." The third suggestion was that the editor not be under the control of the Executive Council in any way, so that the *Journal* would remain free from "any pressure on the part of the Administration that may be in office" through government employees' presence on the council. Despite the depth of the petitioners' dissatisfaction with the state of the *Journal*, they declined to provide any examples of bias or incompetence. Doing so, they wrote, would only lead to arguments and personal recriminations, which they claimed was not their purpose.[31]

The petition, while addressed to the society's Executive Council as a whole, was concerned solely with the *Journal of Forestry*, with no criticism for any other activity or committee of the society. Three weeks after the petition was sent, Franklin Reed met privately with H. H. Chapman at Chapman's summer home, where they tried to come to some understanding of the problem. The petition was sent to all the members of the editorial council but was not sent to Reed, the editor in chief, although the content of the petition concerned him more than anyone else. Reed immediately jumped to the conclusion that he himself was the object of the petitioners' wrath. Not only was his appointment as editor in chief still on an ad hoc basis, but his highly critical review of Marshall's book had been published only six months earlier. However, Reed came to believe that the petitioners were after something more than simply his resignation. Chapman had authored the editorials in the previous three issues, not Reed, so Chapman might be considered more of a motive force for the *Journal*. The presidency of the society, however, was determined by popular vote at the annual meeting, so by definition Chapman must be considered, as much as anyone could be, the voice of the society. If the resignation of neither Reed nor Chapman was the petitioners' underlying desire, the vagueness of the petition's complaints and solutions left the two men unsure of what they were being asked to do.[32] They focused on the particular editorials that might have offended the petitioners and speculated about whether Reed might have mistakenly rejected articles submitted on topics of forest policy and New Deal programs. Also, they tried to guess which particular person the petitioners wanted to oust from the *Journal*. Reed was particularly sensitive to the petitioners' attacks, perhaps because he had not wanted the job of editor in chief in the first place and felt that he was taking the blame for many policies and precedents put in place by his predecessor, Emanuel Fritz.[33]

By August 23, the editorial office of the *Journal* had received 87 replies to the letter of query, which had been sent out on July 18, 1934. While the total number of replies would eventually grow to 137, many respondents worked in the field for long stretches of the summer and had not returned in time to respond by the deadline. The office of the *Journal of Forestry* prepared a mimeographed packet of all the responses, retyped and alphabetized, available to any society member who wished to see the complete contents of the file of responses. Franklin Reed, the editor, prepared a summary and analysis of the responses collected by August 23. He described the range of responses by breaking down the letter of query and the respondents into their component parts, so that if there were certain biases among certain elements of the society's membership, those biases would be apparent. There were no clear trends in the pattern of responses to the letter of query. While more government employees favored changes, and more industrial foresters advocated the status quo, there was plenty of dissension from the norms among each of these groups. Many foresters, in government, industrial, and academic employ, seemed to base their reactions mostly on the imperious and threatening tone of the Zon Petition rather than on the widening rift in the field. There were no patterns in the responses based on the region of the nation where the respondent was employed or educated, nor were there patterns based on the respondent's professional experience. There may have been a pattern based on the political party affiliation of the respondents, although such affiliation could not usually be determined.[34]

Despite the assurances, in private correspondence, that at least some of the petitioners meant no disrespect to Reed and the other members of the council, the atmosphere of suspicion remained at the *Journal*'s offices. In the days leading up to the January annual meeting of the society, this became especially obvious as Marshall tried to use the *Journal*'s facilities to prepare a mailing to all members of the society. Reed confided in a handwritten note to Chapman that "I suspect he wants to circularize something intended to hurt both you + me + maybe the whole Council. If I could trust him to play fair I would be inclined to try to help him out, but I *do not* trust him nor any of his gang."[35] Chapman's response to Reed showed that even though he agreed that Marshall's intentions toward the council were suspect, they should nonetheless comply with his requests for assistance. As mentioned above, Chapman had determined as early as the first part of November that there was little to fear from the petitioners except verbal criticism. Unlike Reed, Chapman had faith that the society's members would be unswayed by the petitioners' repeated attempts to blacken the *Journal's* reputation. "We have nothing to fear from Mr. Marshall," Chapman told Reed. "Every move we have made in connection with giving these men rope has resulted in tightening the noose."[36] Despite Chapman's statement, though, it is clear that Reed was not alone in suspecting that the petition was really a personal attack. Many of the responses to the letter of query also followed this pattern. At the annual meeting, a portion of the evening executive session was given over, at the petitioners' request, to a further discussion of the matter. One of the petitioners, William Sparhawk, was drafted to read a statement from the petitioners and propose further discussion of editorial policy at the *Journal*. As Chapman recalled years later, Sparhawk "came to a realization from contacts with members at the meeting that, to say the least, the petition was unpopular and suggested that he drop it. I told him, 'not on your life. You will present that petition tomorrow morning and abide by its results.'"[37] At the next day's session, just as the discussion was starting to heat up following Sparhawk's presentation, a motion was made to table the matter of editorial policy indefinitely. The motion passed unanimously, signaling to the petitioners that the members of the society did not support them.[38]

Some of the petitioners tried to make it clear that their criticisms were not intended as personal attacks. One of the petitioners, Forest Service forester E. N. Munns, wrote to Fritz that

> apparently you—and others—have conceived the idea that a group of Bolsheviks are on the rampage with war whoops and knives (do Bolsheviks use them?) to take your scalp and that of Reed. You and I have had many differences of opinion, and I hope that we shall continue to have them, over matters pertaining to our profession. You should know that in the present instance, as in others, I am not particularly interested in getting the scalps of any person but of attempting to correct practices which I believe to be not in accord with best public (or professional) welfare.[39]

In their private correspondence other petitioners also tried to reassure the editorial and executive councils that their goal was not the ouster of any individual. L. F. Kneipp wrote to Fritz that it was "only by the accident of current circumstances that the movement [toward changes in the *Journal*] involved Chapman and Reed, two men who command the esteem and approval of all who know them." Kneipp suggested that the assumption by the society members that the petition's intentions were either personal attacks or political statements showed only that they were unable to "reconcile their personal philosophies with the obvious requirements of current and prospective forestry."[40]

Many American foresters had already sensed a rift in the culture of the profession. The tension between the ideals of objectivity and social utility were palpable, even if unarticulated by most professionals. The petition gave form to that rift and supplied a list of foresters who could be deemed responsible for it. Although Marshall was not the most famous or accomplished of the petitioners, many were ready to blame him for inspiring both the petition and the greater rift that its existence implied. Marshall's November critique of the *Journal*'s editorials had only increased this feeling. A letter to the *Journal* by A. E. Wackerman, an industrial forester for the Southern Pine Association who was formerly with the paper manufacturer Crossett and an officer of the society's Gulf States chapter, got right to the heart of a sentiment that was expressed by many. In mid-November, he wrote a letter to Reed that read, in its entirety, "Dear Reed: Must any more of Robert Marshall's stuff appear in the *Journal?* Yours very truly, A. E. Wackerman."[41] The petition, the official responses to it, summaries of the query's results, and counterresponses appeared in the October, November, and December 1934 issues of the *Journal*. While all of this made the rift grow ever more apparent, at the same time it gave the *Journal*'s readers a clearer knowledge of the complexity and diversity of their field. Furthermore, it helped place forestry in historical context by forcing a retrospective view of the profession at the same time it demanded changes in the future. While the petitioners demanded more conservationist thought and more allegiance to the New Deal, these were merely indicators of their real desires. At heart they were demanding that forestry move away from its focus on narrow, applied research goals and move instead toward a new role in American culture as policy maker and steward of forestlands.

Reed, acting editor in chief Chapman, the president of the society, and the rest of the editorial committee and Executive Council responded quickly and decisively to the petition. To determine how widespread this dissatisfaction with the *Journal* was, they sent a copy of the Zon Petition, along with a letter of query of their own, to about 150 members of the society. It is interesting to note that the first action of the editorial staff was to question the membership at large about their opinions. This mimeographed letter of query was written by Reed at the request of Chapman and was mailed out from the Washington offices of the Society of American Foresters on July 18. Most of the letter's recipients had already received copies of the petition during its initial release the previous month, and many had already discussed the matter with their colleagues and come to their own opinions on it. Reed now solicited these opinions en masse in order to gauge the level of support the petitioners had among the membership as a whole. The recipients of the letter were intended to be a sampling of the spectrum of members' backgrounds, philosophies, and employment histories. These recipients also included the officers of all seventeen regional sections of the society, as well as members of the editorial board and the directors of many of the nation's schools of forestry. As Reed put it, the goal was to discover "whether the views expressed in this petition are the views of the majority of the membership, or whether they are simply the opinion of a small minority." Reed reminded the recipients that the *Journal of Forestry* was "the property of the Society as a whole, and of all of its members. Its editorial policies and its methods of management therefore should be such as to meet with the approval of the largest possible majority. It is not the personal property of the Editor-in-Chief, to be run according to his own individual whims and fancies, nor should its policies be dictated by any small minority group or faction." Reed and Chapman, despite their suspicion that the petitioners would not find much support within the rank and file of the society, were nonetheless willing to put the matter up for public debate. The Society of American Foresters differed from other professional scientific societies in its attention to the desires of its members. It saw its primary mission as serving the community of American foresters, not advancing objective scientific research on American forests.[42]

The membership of the society was overwhelmingly against changes in the *Journal* suggested by the petitioners. Many felt that there was already enough discussion of New Deal policies in the *Journal*, and that the petitioners were simply asking for even more focus on such issues. Such foresters thought the *Journal* was, and should remain, mainly a clearinghouse of all sorts of scientific and technical information on forestry. Eighty-seven of the society's members contacted by Reed responded to the *Journal*'s request for comments and reactions to the Zon Petition by the August deadline that he had set. While the *Journal* had participated in a fairly steady level of editorializing and self-analysis over the years, it had done so mainly through the voices of the individuals on the Executive Council of the society and the *Journal*'s editorial council. In this instance they were soliciting the input of members who, while fairly prominent in the field, may never have held office on the national level of the society. In many cases the respondents seemed to relish being asked for their opinion, indeed to relish the machinations of analysis into which the *Journal* had been thrown. It seemed that for the last decade there had been a large amount of unarticulated dissatisfaction with the state of American forestry, resting just below the surface of courteous professionalism.

The petition resonated for many longtime public foresters, who lamented the changes in their profession. For example, James O. (Hap) Hazard, the state forester of Tennessee, wrote that "the *Journal* might almost lead one to think that the prime purpose of present day foresters is research in matters that can have little bearing on the remedying of the present practical conditions which confront American forestry. The forester's job is more than timber growing."[43] However, it would be a mistake to assume all federal foresters held the same favorable opinion of the petition, because some New Deal foresters were against the idea of moving the profession in that direction. For example, H. M. Meloney, a forester for the National Recovery Administration, saw the increasing importance of industrial forestry within the NRA and in the larger realm of American forestry. He cautioned that "while the ideas of this small minority may be of value, to restrict the progress of a profession to their views alone, will have little attraction to types of men who are called upon to carry the responsibilities of industrial management through such trying times as those being experienced now."[44] A number of other respondents echoed Meloney's fear that alienating industrial foresters would in the long run be detrimental both to the Society of American Foresters and ultimately to the forests themselves.

Many foresters disagreed with the assertion that political and social issues should be more prominent parts of the professional discourse. They saw the *Journal* as a purely scientific and technical publication that should maintain a stance of apolitical scientific objectivity. Preservation and social issues belonged in the pages of the *Journal of Forestry*.[45] However, some thought that a new publication might be a good complement to the profession of forestry, as an auxiliary forum for the discussion of political and social issues.[46] Some of the response was based on larger political opinions. Royal S. Kellogg, a Forest Service veteran who had moved into the private sector, described the petition as purely "for purposes of propaganda of pet theories." He noted that all of the signers either were or had been "on the public payroll . . . and the whole crowd are devoted to the idea of practically unlimited expenditure of public money, as might be expected."[47] R. C. Hawley was "emphatically against turning over control . . . to a bunch of radicals."[48] S. R. Black, a California forester, alluded to the new trend toward politically based appointments in the New Deal programs, stating that "to get a job in forestry activities . . . the primary qualification is membership in the Democratic party. . . . It should not be necessary to require that the forestry publications should also become political organizations."[49] Some, particularly industrially employed foresters, seemed to find the petitioners almost comically self-serious, and representative of the worst of the New Deal superciliousness. Charles Boyce of the American Paper and Pulp Association mocked the petition as demanding that "there ought to be some rip-snorting propaganda to arouse a social consciousness as it may be expressed in terms of trees."[50]

Combining the generally accepted goal of scientific objectivity with the activist projects of social and political change was difficult for many foresters to accept. If scientific research was to be done without bias, it would need to remain separate from politically charged causes and activism. Some pointed out that, while important, so-called social forestry was not within the forester's training and that reconciling social forestry with scientific research would be difficult. One forester lamented that "everyone is inclined to save the world by talking social reform, etc., and actually doing nothing. I am not unmindful of the need for certain social changes, but . . . let us not as foresters get so far abroad on the sociology tangent that we lose sight of the fact that we are trained as foresters and that our professional technique should decline by drifting."[51] Indeed, remedying this gap in the training of professional foresters could not be done just by a change in editorial policy but would instead require a fundamental shift in the training and goals of professional forestry.

Indeed, some still saw forestry in a traditional way, more as a craft or a trade than as a science. For them, the profession was best approached as a practice of practical problem solving, in which political issues and land management decisions were not at stake. This sensibility manifested as a sort of defensive apathy, as in one forester's comment that "it was not long ago, that periodically, I became all steamed up about some ideal or other. In retrospect I now see that frequently I was either chasing rainbows or unknowingly helping along some scheme of publicity or propaganda for someone else. I am now content to rely on the uninspired plugging of average men to bring about whatever progress may be justified."[52] Some thus hoped that no underlying political sentiments would exist in scientific foresters. The ideal would then be that despite the utility of their field to economic and political goals, foresters would not stray into decision making on those topics. The use of forest science by political entities, they said, should be completely ignored by those who were actually doing the science. Such responders, interestingly, did not base their arguments on the standard notion of scientific objectivity. They pared down forestry to the point that the profession no longer attempted to make general statements about the forest but rather focused only on the particularities of each case studied.[53]

Indeed, foresters found some value in defining themselves as problem-oriented experts rather than as either impartial scientists or partisan activists. Compared to other professional scientists, foresters were far more engaged with the economic and political trends in the nation. As one forester stated, "I do not agree that the profession has been lax as to its obligations. . . . On the contrary I know of no other profession and no other group which has been as ready to undertake all of the extra obligations imposed on it by the present depression."[54] Calls to ignore politics completely were for the most part the wishful thinking of those who yearned for the hermetic atmosphere of other scientific specialties. These foresters were ignoring the fact that their profession was very different from any other that involved scientific pursuit. The expertise of foresters was singularly important to the continued functioning of the lumber industry, at the time one of the nation's largest natural resource extraction activities and sources of rural employment. Foresters were also pivotal in determining wilderness and park designations and policies for the federal and state governments. And for decades foresters had the ear of politicians to an extent that few other professionals had.[55]

The October 1934 issue of the *Journal of Forestry* devoted eighteen of its pages to the matter of the Zon Petition. That issue's editorial, entitled "Professional Idealism," signed by society president H. H. Chapman, addressed the petition, its criticism, and its relationship to New Deal forestry programs. Chapman indignantly recalled his early work, as far back as 1909, in developing the foundations of what would become the Lumber Code, as proof of his own ties to the New Deal. However, he stressed that the idealistic desire for large-scale preservation needed to be tempered with a realistic acceptance of the necessity of the lumber industry. By accepting that lumbering would go on, foresters could design methods and procedures to ensure a permanent forest resource on the public and private lands of the United States.[56] The "Society Affairs" section of the October 1934 issue reprinted the petition and lengthy explanatory notes by Chapman and Reed.[57] It also included a response to the petition written by Emanuel Fritz, the recently resigned editor in chief. Fritz wrote of the utter impossibility of creating a journal that would ever be able to satisfy its entire readership. He also expressed frustration that the petition was attracting so much attention from foresters, writing, "At a time when forestry has been handed its sweetest victory on a diamond-studded gold platter one group wants to indulge in debate while the rest are straining to practise forestry in the woods. Are we going to let the first group control the *Journal* for its debates or are we going to insist that our official organ provide a proper assortment and an adequate number of technical tools?"[58]

The petitioners wanted less focus on the technical and botanical aspects of forestry, and more engagement with the politics and culture of the New Deal. Furthermore, they were acutely aware that the New Deal offered a singular opportunity for foresters to become major players in national policy debates. The petitioners believed that the only way professional foresters could gain this power, however, was to leave behind the notion of forestry as a scientific pursuit and pursue forestry as a highly engaged, policy-making political enterprise. The concern, then, was not whether forestry was a profession but whether that profession was one of solely academic scientific pursuit or one of a more politically engaged nature. The petitioners were criticizing the membership of forestry because through such criticism they could determine the preexistence of the very form of the profession they desired. Their complaint that the *Journal* was not publishing articles on New Deal forest policy presupposed that such articles were being written by professional foresters but were not being included in the *Journal*. If these articles were not being written, as it seems was the case, then the petitioners' complaint was not with the *Journal* but with the foresters themselves.

Marshall and Zon were aware that the petition would not have much effect but rightly calculated that it would inspire significant introspection in the profession. In fact, Marshall was interested in shaping a new profession, centered on wilderness and social causes. It would not only be separated from any industrial influence but would in fact fight against the very influence other foresters were welcoming in. The petition stated that "one of the proposals which has been crystallizing, is the publication of an independent organ to fill the gap left by the recent purposeless editorial policy of the *Journal*. . . . We are not unmindful, however, of the possible effect of such publication upon the official organ of the Society and upon harmony within the Society itself."[59] In fact, splitting off to create a new journal had been the coterie's original intention, and they developed the idea of a petition only later. One of the signers, W. C. Lowdermilk, described this to Franklin Reed in a meeting on August 15, 1934. Lowdermilk was a member of the *Journal*'s editorial council and the director of the Bureau of Soil Erosion Control in the Department of the Interior. He told Reed that when Zon had initially approached him, the "first proposal was to start a new publication right away but that [Lowdermilk] argued against it" and urged them to try to change the existing structures of the discipline instead.[60] The petition was thus calculated as a provocation and a parting shot. Lowdermilk himself made it clear that he did not favor the creation of a new journal, and that he had not been one of the most militant of the petitioners. He had signed the petition mainly because of its demand to adopt a more open editorial policy. Indeed, Lowdermilk was still smarting from the rejections of his work on erosion control and forests under the previous editor in chief, Emanuel Fritz. While his work had been scientifically sound and technically oriented, it had not fit into Fritz's editorial criteria. Lowdermilk's interdisciplinary approach to integrating forest and soil work had been deemed too far from the standard to merit publication in the journal. His interest in signing the petition was mainly in ensuring that the technical articles in the *Journal* were of a wider scope, not in supplanting technical articles with policy briefs.

If Lowdermilk represented the mildest motivations of the petitioners, wishing only to modify the current *Journal* policy, then Bob Marshall, with whom Reed met the next day, represented the opposite end of the petitioner spectrum. For Marshall, the primary goal of the petitioners was not just to change the *Journal* but to create a completely new publication. Reed suspected that the only way Marshall could be satisfied with the *Journal of Forestry* was if there was a complete turnaround in its policies; if, as Reed wrote, there was "an Editor who will write a series of ringing editorials expounding and supporting his (Marshall's) forestry and social theories."[61] If the majority of the society's members did not agree with the petitioners on these changes in editorial policy, Marshall told Reed, then it of course would not come to pass. But if that was the case, then "he and the group thinking along with him would have to get busy and get out their own publication."[62] In this conversation with Reed, Marshall had not purported to speak for all the other petitioners, since, as Lowdermilk had already made clear, not all of them were willing to go along with a new publication. The threat of a new publication was not idle, however, and funds for developing it were indeed available. Reed, along with many others, suspected that the source of these funds was Marshall's own deep pockets.

The petitioners wanted a shift among foresters from thinking of themselves as scientists who studied the forest to thinking of themselves as prophets of forest destruction and heralds of change. The Zon coterie was anxious to imbue forestry with a sense of social purpose and New Deal utility. Personal and political alliances had much to do with the petition. Zon's and Pinchot's long histories of involvement with the *Journal* and the society prompted them to see the changes in forestry as something that could be guided by changes in the institutions that delineated the profession. The content of journal articles was not the problem—it was the editorial tone and overall sentiment of the profession that were really the heart of the complaint. The *Journal* was already covering a wide range of forest-related issues in its pages, including those issues of interest to the petitioners. Given the nature of the *Journal* at the time, the arrival of the Zon Petition might have seemed surprising to Reed, Chapman, and the editorial staff. In the months since Roosevelt's inauguration, the *Journal* had printed a recurring series of articles on the designing of the Lumber Code of the National Recovery Administration as well as articles on a number of other New Deal forestry initiatives.[63]

Members of the coterie were notable for their radical politics, but they also stood out from the norm in other ways. The group included many who did not fit into the classic, middle-class Anglo-Saxon mold of the professional foresters of the time. Marshall and Zon were both Jewish, and in addition, Zon was a Russian émigré. Pinchot and Ahern, although they had once held prominent positions, had spent more recent years lobbing criticism at working foresters from outside the profession. The cultural backgrounds of these men branded them as outsiders to the rank and file of professional foresters. One can speculate that their unique outsider perspective on forestry led them to develop a piercingly critical perspective on their own profession. Their outsider status did not get in the way of their advancement within the profession, as both men held prominent government posts. It did, however, contribute to the negative reaction to their petition and fueled some of the vitriolic responses to it.[64]

The perceived outsider status of Marshall, Zon, and by association the rest of the petitioners was accompanied by a deep suspicion about their political affiliations. Marshall was known to be a socialist and the president of the Washington chapter of the ACLU. Pinchot, despite beginning his career as a Republican, had proved to be an enthusiastically radical New Dealer as a governor. Zon's radical political past was less well known, but for many his Russian origins linked him to the leftism of the new Soviet Union. Many assumed that the petitioners' real motive was to radicalize the *Journal* and the profession. Indeed, their propositions both within the petition and in the larger political sphere were propositions to make the United States more socialistic in its policies. But for many of the society's members, this in itself was a reason to hate the petitioners. Some foresters, both publicly and privately employed, were just waiting out the New Deal with the hope that the country would return to a more conservative forest policy once the national economy recovered. One industrial forester wrote to H. H. Chapman that "here then is the crux of the [petition's] whole argument—*socialization* of the forests. . . . The majority feels that socialism in any field is Un-American and unwarranted. It is patterned after Stalin-ism, or Fascism."[65] Hatred for the petitioners' politics was often accompanied by a feeling that the petitioners were interlopers who had no place in American politics or American forestry. Said one forester, "I am so against anyone who advocates this so called New Deal that I can't say a good word for them. The whole damned thing is a political racket and if allowed to go on will wreck the country."[66] In fact, the petitioners actually did form a social and professional network outside the normal circles of the profession. A pattern of xenophobia and anti-Semitism in the society members' responses suggested further that while they were near the top of their profession, as delineated by the content of the *Journal* and by the society's membership, Zon, Marshall, and their allies were still unwelcome outsiders.

The editorial board of the *Journal* clearly wished to be done with the matter of the Zon Petition completely. A brief note announcing changes in the *Journal* in advance of the appointment of a new editor in chief, Herbert A. Smith, appeared in the January 1935 issue. The note, which appeared above an editorial signed by Reed and Chapman, stated that "the editorial policy will continue unchanged. It can perhaps best be expressed by quoting from our first editor-in-chief [Raphael Zon] in his letter of resignation."[67] The note's quotation from Zon's 1928 resignation letter was his desire for the future of the *Journal* that "no creed, however extreme, no theory, however radical, should be barred from the pages of the *Journal* just because of the ideas expounded. . . . And no editor, whether he is a school man or a Forest Service man, should ever use the *Journal* for his personal glorification of the institution to which he belongs, but always maintain it as an independent, common forum for the entire society."[68] In recalling the words of Zon during his last days as editor of the *Journal*, the writers were both recalling the original purpose of the *Journal* and acknowledging that the petitioners' complaints might have indeed had some merit. However, the *Journal* did not mention the Zon Petition specifically, and the issues under the new editor in chief contained articles of the same spectrum of opinion as before. The editorials, if anything, might have been even sharper and more critical toward the ideals of New Deal forestry.

Post-Forestry: New Directions for "Wilderness Science"

The petitioners sent a pamphlet to most of the members of the society in January 1935. This second missive clearly did not concern changes in the *Journal of Forestry* but instead profoundly criticized the discipline of forestry itself. The petitioners' policy demands were the same as before, but this time they stated essentially that anyone who did not agree with them on policy issues was not actually a scientific forester. Anyone who correctly understood the scientific bases and recent findings of forestry, they stated, could only come to the same conclusions as the petitioners. They cited recent scientific findings on forest ecology and the economics of logging yields and asserted that the only logical conclusion of a scientist who examined those findings would be the same as theirs. And conversely, if a forester disagreed with their policy beliefs, the pamphlet asserted that such a person then did not accept the scientific findings that they said underlay their policy beliefs. The editors of the *Journal*, and foresters in general, paid little heed to the January missive. This pamphlet, with its elitist and insulting tone, hardened the rift in the profession and did nothing to sway those who disagreed with its writers.[69]

Even among those who still agreed with the petition, the solidarity from the summer of 1934 did not carry over past the winter of 1935. The petitioners had signed on for very different reasons, congruent with their very different places within the discipline of forestry. They did not all feel equally strongly about every aspect of the petition, and they were not all equally adamant about the changes they demanded. Pinchot had turned his attention to other matters, and Zon lamented that he had not heard from the old man in quite some time. Zon himself was beginning to grow weary of Marshall's constant sparring with the Society of American Foresters. As Marshall turned his energies toward wilderness preservation, Zon, who was driven more by desire for scientific rigor, began to step back from the fray. Using his own funds, Marshall founded the Wilderness Society in the spring of 1935 to promote the ideas and policies that the society had refused to support. The main goal of this new society was to advocate the preservation of all undamaged American lands as untouched wilderness under public ownership. The Wilderness Society was not a scientific organization, instead bringing scientifically trained men like Marshall and Aldo Leopold together with policy makers and public relations experts. One of the express goals of the Wilderness Society was to use scientific understanding of the forest to further political and social goals for the preservation of forests and other wild lands. Thus under the aegis of the Wilderness Society, its scientifically trained members could use data from any scientific discipline as it suited their needs, while being beholden to none of them. Professional foresters, on the other hand, could draw conclusions and make recommendations only within the narrow confines of their regional and topical scientific specialties. Furthermore, the Wilderness Society did not require its members to leave behind the standards of their own disciplines but allowed them to separate their scientific conclusions from their policy conclusions. Marshall remained highly critical of the growing industrial focus of forestry, in letters and journal articles as well as in the Wilderness Society's magazine, *The Living Wilderness.*[70]

Journal of Forestry editorials continued to be critical of Marshall and his remaining allies, most notably in an editorial critical of the Wilderness Society, published in December 1935. Entitled "The Cult of the Wilderness," the editorial was the work of the new editor in chief, Herbert A. Smith. It characterized the aim of the Wilderness Society as based "on a questionable assumption . . . that the primitive wilderness possesses an exquisiteness and richness of beauty which, once lost, can never be recovered, since it is the perfect and precious flowering of the ages, possible only where continuity has never been interrupted."[71] It pointed to the elitism inherent in the twentieth-century vision of wilderness, whereby only those people not engaged in work, but rather at leisure, may visit the wilderness. The editorial also contrasted the Wilderness Society's model of the forest with the model used by foresters and other scientists who assessed landscapes based not on subjective notions of beauty but on quantifiable, presumably objective, factors. Assessing the value of a forest, the editorial stated, "is to be sought not in transcendental or pantheistic nature idolatry but in our good old fashioned doctrine of obtaining, through wise use, the largest measure of contribution to the welfare of everybody in the long run. . . . To make it the object of a cult is as irrational as to mourn the vanishing of the vast herds of buffalo before the advancing tide of human occupation of the prairie grasslands."[72] Such opposition to the Wilderness Society, and to other manifestations of the ideals of the Zon Petition's signatories, was common in the *Journal* during the years following the petition. It points to the widening of the gap between the two groups, as foresters increasingly took positions in opposition to those of preservation advocates.

Marshall founded the Wilderness Society not just because he wanted a platform for promoting his beliefs but because he had been denied that platform by the Society of American Foresters. The Wilderness Society, founded in the immediate aftermath of the contentious Zon Petition, was not offering an alternative model of a profession. Its goal was to use elements of forestry in wilderness advocacy, thereby channeling existing knowledge in a new direction. When the Society of American Foresters declined to engage more than necessary with New Deal politics and advocacy, Marshall created his own outlet. Marshall, Zon, and some of the other signatories had heralded the New Deal as an opportunity for professional forestry to adopt a new spirit of advocacy. The Wilderness Society was created as a forum where scientifically trained foresters could use their training in pursuit of another goal, wilderness preservation. The Society of American Foresters had made it clear that professional forestry had no place for such advocacy, and the Wilderness Society was formed in reaction to that decision. The conservative response to the Zon Petition inspired Marshall to design his new society with a more radical agenda, at least in part as a reaction to that response. The Society of American Foresters attributed its own conservativeness to the constraints of scientific detachment, which led Marshall to question the need for scientific credentials in an advocacy group at all. The Wilderness Society required no scientific credentials from its members and deemphasized scientific information in relation to wilderness advocacy and appreciation.[73]

The coterie recruited the well-known conservationist Aldo Leopold as a founding member of the Wilderness Society. Leopold, trained as a forester at Yale, spent the first part of his career in the Forest Service. After beginning in the Southwest, he spent time at the agency's Forest Products Laboratory in Madison, Wisconsin. He took a position as a professor of game management at the University of Wisconsin in 1933, where he developed theories of wildlife ecology and wrote on conservation issues for the *Journal of Forestry* and other outlets. Marshall contacted Leopold with the proposition of coming on board with the Wilderness Society. Having just published his 1933 textbook, *Game Management*, which laid out a way to use ecological science to manage land for primarily recreational use, Leopold was a herald of a new type of conservation. He showed how existing forms of science and management could be infused with new life through the contributions of ecology, genetics, and other new scientific disciplines. Leopold had known Zon for a long time and respected him as a researcher and as a person. He had praised him for "that quality of your mind which has aided so much to the intellectual timbre of our profession," and for showing other foresters "how to think joyfully, which is usually the same as thinking well."[74] However, Leopold had declined offers to be involved in the coterie's previous critiques of the profession, despite affinity toward their philosophy. He maintained a studied neutrality, observing that "only a few . . . have expressed themselves in the *Journal* in recent years on these burning questions of social policy. . . . I thoroughly agree with them that history is being made every day, and that there are a multitude of important questions, especially in social policy, which need discussion but are not being discussed. I thoroughly agree that we are ossifying, but it is up to the members as well as the editors to re-vitalize professional thought."[75] Leopold foresaw potential radical future changes in the field but thought that battling over definitions was secondary to the real work of conservation.[76]

Conclusion

The Zon Petition stands as an interesting challenge to the standard notion of scientific discipline formation. It is a singular example of a coalition of members of a scientific discipline demanding, from within the ranks, that the discipline not continue as such. Forestry defined itself as scientific, although it differed in many ways from other, narrower disciplines. The petition pointed to some of the weak points in the makeup of American forestry and led the membership of the society to question the functionality and effectiveness of their field. The utility of forestry science to the conservation policies of the New Deal meant that for the first time forestry could have a lot of power in American culture. The petitioners found this power to be more important than the scientific trappings of the discipline, and their attempts to change the field were a result of this. They challenged the members of the discipline to leave behind their conception of their own field and take on this greater role in the political and cultural fabric of the nation. With increasing power, however, the foresters would leave behind the ideals of objectivity and segregation that were part of the discipline. Most foresters rejected this call to national prominence, however, preferring to remain a secure if less powerful scientific discipline. The episode thus serves as a challenge to the notion that disciplines, once formed, remain static, and to the notion that increases in scientific rigor are always welcomed by members of a given field.

Many of the foresters who responded to the petition, among both those first queried and those who reacted after seeing the published accounts in the October issue of the *Journal*, already sensed a rift in the culture of American forestry. The debate showed foresters how intellectually diverse their profession had become. On the face of it, the petition had simply asked the *Journal* to include more favorable coverage of conservation and of New Deal programs. However, readers understood that these demands were merely indications of deeper discomfort. The coterie wanted forestry to redefine itself, to acknowledge the human presence in the ecological system and to consider foresters' roles as stewards of American forestlands. For foresters in the Northwest and throughout the nation, these demands threw into high relief the differing motivations and philosophical tendencies of those members of the profession working on clear-cut reforestation, national forest management, and wilderness preservation.

5 The Money Tree
Private Forests and Public Relations

Lumbermen watched the furiously idealistic New Deal turmoil in the ranks of professional forestry and wondered how it would play out on the ground level in the Douglas fir region. Professional foresters' struggles over the intellectual and political aspects of their profession in the early 1930s had originally been sparked by concern over the rate and destructiveness of new western logging operations in the early twentieth century. While radical foresters stated their aim to keep the nation's forests from complete decimation, the introspective debate instigated by Raphael Zon and his allies often seemed to be more about the meeting rooms of Washington and the lecture halls of academe. These intellectual disagreements about the true aims and best practices of their profession would, in the end, be crystallized in the management of actual forestlands. Changes in how foresters conceived of the forest could restrict or expand Douglas fir logging through regulation in the Northwest. Likewise, dramatic shifts in forest access could occur if public sentiment about forest use was swayed by the impassioned rhetoric of activist foresters like Bob Marshall or Aldo Leopold.

From their Pacific Northwest vantage point, lumber company executives and regional forest managers watched the drama unfold back East. Changes in federal forest policy enacted in Washington affected lumber activity in the Douglas fir; hence news about the profession could be vitally important to profits. Journals, correspondence, and conferences provided news not just of technical advances but of changes in professional attitudes. The Portland-based magazine *The Timberman* and other lumber trade journals relayed the wrangling within the Forest Service and elsewhere in the state and national governments. The annual Pacific Logging Congress provided a place for face-to-face exchange and discussion between loggers and the Forest Service. Such venues were places to learn not just of new technical advances in logging and milling but also of news and rumors from Washington.[1] As we saw in the previous chapter, lumbermen with professional backgrounds in forestry often read, and at times wrote for, the academic *Journal of Forestry*. *American Forests and Forest Life*, the journal of the American Forestry Association, published the latest news of the profession for a nonacademic audience. Northwest lumber executives kept up with this news not just to stay abreast of technical breakthroughs but also to learn how

Logs coming up a conveyor into the Weyerhaeuser sawmill, on the banks of the Columbia River at Longview, Washington (1936). Library of Congress, LC-USF34-004837-D.

the profession regarded their enterprise. Corporate decisions could hinge on executives' predictions of future state and national regulatory changes.

Many whose businesses depended on access to the Douglas fir understood the industry's precarious perch in public opinion. The first decades of the Forest Service's existence had proven the importance of a favorable regulatory environment to corporate profits.[2] Endeavoring to keep federal foresters and other government regulators industry friendly became important for logging companies. Lumber trade groups like the West Coast Lumbermen's Association (WCLA) monitored actual and potential changes in forest policy that could affect forest ownership, logging, and wood products marketing. The lumbermen developed ways to affect Washington policy making directly, through lobbyists like the WCLA's Wilson Compton or through lumber-friendly politicians. On the regional level, too, political influence proved valuable in creating favorable taxation structure and limiting onerous land-use regulation. However, political allies were valuable only up to a point because politicians were subject to the whim of an even more potent force: the opinion of the voting public. Lumbermen thus began to work on finding ways to cast their companies' treatment of forests in a positive light, with the hope that it would win them support from professional foresters, policy makers, and the public at large.

This chapter focuses on changes in the ground-level management of the Douglas fir forests, and how criticism and debate among foresters influenced the direction of those changes. The new directions, although impacted by the turmoil within the profession, were far from what Zon and his colleagues wanted. Instead of embracing an ecological view of forest function, these forest managers focused on the "money tree" found in them, the Douglas fir. This chapter begins by examining the hiring of trained foresters by American lumber companies as a way to both improve productivity and create a positive public impression. These privately employed foresters described themselves as practicing industrial forestry, keeping an eye on the bottom line while acting as problem solvers for their employers. The abstract ideal of sustained-yield forest management became a real agenda for public forestry, intensifying the focus on the money tree at the expense of a wider view of forest health. Public lands sustained-yield management began in 1937, when parts of the Oregon Douglas fir forest were covered by new legislation governing logging in the revested railroad land grants known as the O&C lands. Lumber executives worried they would soon be compelled to submit to highly regulated sustained-yield management on their own lands. Industry public relations depicted forest management on private lands as both scientific and sustainable. The aim with these campaigns was to find a less onerous alternative to sustained-yield management, to mollify a concerned public, and to minimize further criticism from mainstream professional forestry. Mainstream foresters did push back against these industry moves, even as they found their profession increasingly focusing on meeting the needs and goals of lumber corporations. Consumed with projects outside the realm of forestry, especially the creation of the Wilderness Society, the radical foresters of the Zon coterie made less protest than they might have in years previous. With the sudden deaths of Bob Marshall and Ferdinand Silcox in 1939, the battle-weary radicals weakened even more.

Industrial Forestry and Its Place in the Profession

Logging Douglas fir could be lucrative, but it was a precarious undertaking from a long-term business perspective. Executives knew, of course, that the supply of virgin timber in the United States could not last forever because the rate of cut had become unsustainable by the 1920s. Long-term health for any lumber company dependent on the Douglas fir meant expanding the company in some way. One option was to diversify, investing in new industries and resources. Some lumber companies developed their landholdings by leasing acreage for mines or orchards. Many created separate real estate companies to sell parcels of their cutover lands for farms and settlement. Some lumbermen made a fortune logging and milling, only to dismantle their companies as profits dwindled, moving their money into new ventures far from the forest. Oregon and Washington bemoaned a blight of small operators who intentionally bankrupted their own companies, extracting the profits and leaving behind debtors and clear-cuts.[3] The Weyerhaeuser family chose not to leave the industry that had made them their fortune, instead expanding the scope of their lumber enterprise. In the first decades of the twentieth century, Weyerhaeuser expanded to encompass operations in almost every aspect of the lumber trade, from timberlands to steamship transport to consumer paper. Its unique commitment to the forest seemed to have as much to do with the family's heritage as it did with good business. During these decades, most of the company stakeholders were members of the founding family. While some managers had been brought in from the outside, most of the high-level executives of the company in the 1930s were members of either the Weyerhaeuser family or the families of Frederick Weyerhaeuser's original partners in Illinois and Minnesota. Weyerhaeuser was self-financed, not traded publicly, so there were no stockholders to please, and the company did not need to see profits accrue quickly for the benefit of short-term dividends. The Weyerhaeusers plotted the course of their family business by favoring stability and consistency over diversification, speedy growth, or short-term profits.[4]

Weyerhaeuser had been the first large American lumber company to consider feasible plans for ensuring permanency of lumber supplies through implementing forestry methods on its lands. As early as 1910, Weyerhaeuser representatives began to voice opinions about forestry that would eventually transform industrial forest management. George S. Long, who spent most of his career managing various Weyerhaeuser operations, described Weyerhaeuser as "believ[ing] that the only way in which the forests are likely to be replaced is for the state either to buy the lands from the lumber companies at a small price and replant the cut-over areas or remit the taxes."[5] The company had a long-term interest in both ensuring the reliable flow of timber and minimizing its tax burden. If cutover land taxes were eliminated, the corporations could keep ownership of such lands indefinitely, while investing little time or energy into their maintenance. While Long knew that natural regeneration would be a much slower process than artificial reforestation, with proper taxation and forest management he predicted eventual renewal. As it stood in 1910, private lands logging was untenable for multiple harvests because "it is a simple mathematical demonstration that it will not pay the lumber corporations to keep up tax payments and wait for a new crop of trees on cutover lands."[6] Unfortunately, as discussed in an earlier chapter, foresters would spend the next two decades uncovering the difficulties of Douglas fir reforestation, showing that Douglas fir could not be expected to regenerate naturally even at the slow pace Long anticipated. Weyerhaeuser's primary long-term goal would always be to make a healthy profit, hence its halfway position on reforesting cutovers. Its investment in silviculture would never extend beyond what it anticipated to have as financial payback under the right conditions. As discussed in previous chapters, forest research and regeneration failures both demonstrated that simply holding on to the land indefinitely would not ensure a second harvest of Douglas fir. Artificial reforestation was significantly more expensive than natural regeneration, however. Only after a string of decisive failures of natural Douglas fir regeneration did commercial interests consider investing the money and personnel required to implement artificial reforestation. Federal and state tax codes and regulations evolved in ways that encouraged lumber companies to keep ownership of lands and invest in regeneration, but many were still unconvinced of the economic viability of private industry reforestation programs.[7]

In the late 1920s, a new phrase entered the lexicon of American foresters and lumbermen: industrial forestry. If no scientific forester was employed or consulted, the forest management decisions for a lumber company might be undertaken by managers without any forestry training, who based their decisions on tradecraft and tradition at best, and on convenience and cost at worst. The new industrial foresters intended to replace this hodgepodge with the systematic use of scientific forestry in industrial decision making. Many of the original industrial foresters were consultants who had come into this specialty late in their careers. There were few entries into professional employment besides federal and state employ in the 1920s, but some foresters who had proven themselves within the ranks of government forestry moved on to consulting. Towns, schools, estate owners, and other private landowners hired consulting foresters to survey and advise them on their forested landholdings. The consulting forester would construct management plans and oversee logging operations or other activities that required professional expertise. Consulting foresters' job skills were modified from their previous training in government or acquired from working partners, not gained through formal education. Forestry schools trained people to meet the needs of public forest management, since this was the employment most forestry students aimed to land. By the end of the decade, it was estimated that twenty companies in the western region used the expertise of trained foresters at least on occasion.[8] While many of these were consulting foresters, as the work became more specialized and more tailored to the needs of large corporations, the term "industrial forester" appeared.

Companies hired industrial foresters ostensibly to make sure the land recovered from logging and would eventually provide a second growth of trees. As they hired these consultants, lumber companies also publicized their use of so-called industrial foresters to ensure long-term health of their lands. Regulators, politicians, and the public began to question whether particular companies were "practicing forestry" on the lands they were logging.[9] Landscapes in the Douglas fir region could be permanently damaged by the process of logging, as a result of topsoil erosion, clogged streams, and permanent loss of forest cover. The claim of "practicing forestry" implied that logging operations took into account the limits of the landscape to recover from logging, that they employed rigorous oversight, and that they made provisions for the future health of the landscape. Industrial foresters would ensure the recovery of the landscape following clear-cutting by leaving seed trees, clearing slash, preventing fires on denuded land, and making other similar provisions for future growth of new trees and health of the landscape. Academic foresters questioned whether industrial forestry was really forestry at all, in some cases simply because they were loath to give the industry this sort of professional validation. The so-called industrial forestry, claimed many, was nothing more than a feeble attempt to minimize damage in the hope of avoiding regulation. The apparent motivation for lumber companies to publicize their employment of foresters seemed proof enough to some that corporate motives were corrupt.

At the National Commercial Forestry Congress in 1927, a group of foresters employed by lumber companies argued that professional foresters belonged in industry—that the discipline held the potential to transform the logging industry. Raphael Zon, in his role as editor of the *Journal of Forestry*, reflected on the message of the National Commercial Forestry Congress:

> Industrial forestry, in its simplest terms, means nothing else but— *treating timber growing as a business enterprise.* There is more substance to this definition than may appear on the surface. The admission that *timber growing* may be considered a business enterprise is a new, revolutionary idea. A few years ago anyone who would mention timber growing as a business would be laughed at. Timber growing as a mere fad, to be engaged in by some land owner from an altruistic or patriotic motive, as repentance for past destruction, *yes*; but as a business, *no*. For the first time in our history, the lumbermen as a group begin to look upon timber growing as a business enterprise, as an industrial pursuit, as something that may yield profit.[10]

While many fields of professional forestry involved helping lumbermen profit, silviculture had not up to that point. To academic foresters, especially radicals like Zon, forestry centered on the reparative and experimental processes of understanding and managing the health of growing forests. While loggers had availed themselves of professional foresters' expertise in logging technology, road building, timber assessment, and so on, silviculture had not been their concern. The new industrial interest in silviculture, as Zon pointed out, changed the purpose of the endeavor. Instead of repairing damage or afforesting barren land, the ultimate goal of tree planting for these industrialists was to generate profits. While rigorously planned reforestation would still be beneficial to clear-cut areas, the new industrial forestry displaced some of the profession's idealism.

A critical chorus of professional foresters found many reasons to disapprove of industrial forestry. Most observers were reluctant to define the actual reach of forestry into lumber companies themselves based solely on the companies' installation of a forester on their employee payroll. Thornton Munger reported to his superiors in Washington, DC, that among those operating in the Douglas fir region "I think it is misleading to say that there are six, eight, or ten companies doing thus and so, when there are hundreds of companies in the region all doing a gradated amount of this and that."[11] The bits and pieces of work deemed "practicing forestry" were not enough to satisfy Forest Service standards. Furthermore, in most cases government foresters still managed the industry's complaints and mistakes in fire protection, insect damage, tree disease, and tax reform. The government foresters did all of the hard work, while company foresters reaped the benefits. Others recalled the long history of the lumber industry's antagonism toward professional foresters' regulation of logging activity. H. H. Chapman of the Yale school complained that "it is obvious that [lumber companies] have not yet absorbed the first principles of what constitutes the business of forestry. . . . They stubbornly refused to take any chances, preferring increased profits and forest destruction to an investment which . . . they did not wish to take."[12] For Chapman, as with Zon, there was a deeper problem in accepting the concept of industrial forestry: how to reconcile capitalism with environmental concerns. The lumber industry was focused on maximizing profits in the near term, and it was impossible to maximize those profits while at the same time practicing forestry on its lands. On public land, the long-term interests of the citizenry meant that there would always be a need for prudent use of forest resources. The national forests had been instituted in part to ensure lumber resources for the foreseeable future. But on private lands the story was different. If there was no reason to invest in long-term availability of lumber, Chapman doubted that there would ever be motivation for the quality of forestry to improve on private lands.[13]

However, some foresters worried that outright rejection of industrial forestry would simply alienate the industry, unwittingly making stability of forest resources more difficult to achieve further down the road. Instead, perhaps industrial forestry could be molded to fit the profession's ideals. L. F. Kneipp, a forester employed by the US Forest Service, postulated that "industrial forestry has to be considered from the viewpoint of a general manager under strict obligation to his stockholders to safeguard their investments and make them financially productive, or from an investor who is being asked to put his money into a forestry enterprise. . . . It is, after all, the attitude of the private investor that is going to determine the place of industrial forestry in our future forest economy."[14] A true industrial forester, to Kneipp, would be one who did not focus simply on solving problems related to that company's production of lumber but would also take into account the ancillary effects of the logging process as part of his or her responsibility. Public foresters must pay attention to the "indirect returns to society in the way of better environment, stabilized streamflow and climatic conditions,"[15] and hence so should those who were hired as industrial foresters. Kneipp, like the industrial foresters themselves, believed that creating a successful industrial forestry would require reformulating taxation and regulation to stop penalizing companies that did not abandon cutover lands. However, he also suggested that a real, full-fledged industrial forestry would require ceasing all public subsidy of the lumber industry, so that logging the public timber supply would cost companies the same amount as logging purchased or leased private lands. He worried that if for-profit industry were not separated from public forest management, foresters working within the industrial setting would always be hobbled by the demands of the industry itself. Such unease with the marriage of industry and forestry was typical of the attitude of the rank-and-file federal foresters, who were used to thinking of the lumber industry as an outside force.

Industrial Foresters, Sustained-Yield Forest Management, and the O&C Lands

A Berkeley professor turned Portland-based industrial consulting forester, David T. Mason, has been widely credited with developing the first plan for American sustained-yield forest management as a way to check habitual overharvesting threatening to decimate the Douglas fir.[16] By 1927, easy access to western timber, improved logging techniques, an overabundance of sawmills, a lack of regulation, and an improved distribution network had led to an enormous supply of lumber in American markets. Douglas fir loggers were in danger of going down the path that the upper Midwest's loggers had followed in a previous generation. While those woes affected the industry nationwide in the late 1920s, they were felt especially severely in the fast-growing Douglas fir market. Market supply of sawn Douglas fir had surged far ahead of worldwide demand for the timber, and the prospect for the forest's future health looked grim. In earlier decades, companies had tried solving overproduction by forcing limits on sawmill capacity, but closing individual mills had little effect on the market and occasionally provoked significant labor unrest. Trade extension had amplified demand for Douglas fir lumber, and companies increased distribution and advertising, but it had not been enough to stabilize the market. An attempted merger of forty Northwest lumber companies failed in 1926, largely because of Weyerhaeuser's unwillingness to allow it. Despite this earlier failure, the regional industry began to cooperate by empowering William B. Greeley to increase the marketing and lobbying efforts of the West Coast Lumbermen's Association. However, despite increased unity of purpose among the region's lumber producers, loggers continued to overharvest, and sawmills continued to overproduce, confounding efforts at stabilization.[17]

In this atmosphere, Mason, his business partners Carl M. Stevens and Donald Bruce, and a handful of other industrial foresters and lumber executives began formulating a new concept for stabilizing the American lumber industry, which they termed "sustained yield forest management." The idea of sustained yield has always been part of natural resource economic theory, not just for forests but also for resources like fish and game. Under various names, sustained-yield management had been used by European and Japanese foresters long before the 1920s.[18] Much of the early rhetoric of Forest Service conservation also hinted at sustained yield by advocating the curtailing of logging to maintain a healthy forest. Sustained-yield forestry should not be confused with the more recent term, sustainable forestry, which carries connotations of not just stable forest growth but also stable communities and environmental well-being. The initial era of sustained-yield forest management in the United States was the product of an industrial mind-set in which success was measured by economic stability and consistent profit. Sustained-yield plans, whether formulated by publicly or privately employed foresters, cast the forest as a generator of wood rather than a natural environment. Sustained-yield plans were always a simplification, and usually an oversimplification, of the ecological realities of a forest in time. Foresters and economists had to regularly overlook variables in geography, assume constant rates of growth, and underestimate fire risk to create workable sustained-yield plans. Even without much proof of its efficacy, by midcentury federal foresters' use of sustained yield abounded largely because of the efforts of the industrial foresters working in the 1920s and 1930s.[19]

While sustained-yield forest management proponents publicly highlighted its contributions to the long-term public good, privately they observed that it might also benefit the corporate bottom line. A region-wide sustained-yield policy could serve as a method for corporations to force a curtailed rate of logging, which could stem the oversupply and hence benefit lumber producers. Perhaps the most effective way to correct the American lumber industry's financial woes would be to use sustained-yield forest management as a means to, as David T. Mason put it, "convert the chronic buyers' market into a chronic sellers' market." Mason, perhaps the region's most fervent proponent of sustained yield, observed that it was essentially a restriction of supply carried farther up the production stream by restricting the number of logs felled in the first place. In a characteristically ambitious letter to one midwestern lumber company executive, Mason observed that while curtailment of supply was important, "to be effective the curtailment must be carried back into the forest, and adjusted to forest production capacity; this is the sustained yield plan; and this is in harmony with both industrial and public welfare."[20] In rough calculations Mason estimated that "we can no doubt go on over-producing and flooding the market for at least twenty years," after which the forests would be exhausted and the industry would go bust. On the other hand, "we can apply sustained yield in a way to curb production effectively well within ten years," followed by an almost unlimited future of controlled scarcity in the market. Taking into account taxes, land prices, distribution costs, and so on, he believed that the costs inherent in switching to sustained yield would be far outweighed by the benefits of setting higher consumer prices for scarcer lumber. He believed that nationwide, if lumber companies "continue overproduction, we shall doubtless get something for the timber when it is cut, but what we get will probably be at least $750,000,000 less than what we would get if we took the trouble to install a system which will curb production and give us a sellers' market."[21]

The flaw in the sustained-yield plans industrial foresters conceived in the late 1920s was in how to, as Mason put it, "install" harvesting restrictions. Sustained-yield management of any resource depends on all harvesters of that resource complying with agreed-upon harvest limits. If Douglas fir logs were being harvested for a hodgepodge of different lumber companies and independent contractors, and if there were no significant difference among the logs harvested by different loggers, then restricting the flow of this product from certain loggers would just improve the profit for those who were not complying with the restriction. If some subset of harvesters did not comply, the compliant parties would lose money in the short term. Also, the sustained-yield plan required lumber companies to put their faith in the industrial foresters' mostly unproven ability to predict growth rates

*Small-scale salvage logging operation, colloquially termed "gyppo" logging, in the
forests of Tillamook County, Oregon (1941). The small scale and financial instability of
such operations meant many evaded regulation, leading both industry and government
foresters to blame the "gyppos" for wildfires and other forest management problems.
Library of Congress, LC-USF33-013201-M1.*

of Douglas fir forest. The foresters would be tasked with monitoring the growth
and health of all private forest tracts, and to mandate adjustments of harvests if
the forest were not growing as expected. For sustained-yield management to be
successful, then, a vast majority of lumber companies would have to put their
collective faith in it, voluntarily commit to a common program, and then allow
their lands and activities to be intensively monitored. In 1929 several lumbermen
floated a plan to appoint William B. Greeley, who had recently stepped down as
chief of the US Forest Service, as director of lumber activity in the Douglas fir
region. Greeley, who was a stalwart industry friend, would have been empowered
as a dictator over the activities of lumbermen and the monitoring of forests. While
some sentiment existed that "the industry needs a Mussolini,"[22] in actuality even
if the lumber companies had agreed to such a scheme it would have undoubtedly
violated numerous antitrust laws. Greeley would instead take a job as the
secretary-manager of the West Coast Lumbermen's Association, a major lumber
industry trade group.[23]

Mason's consulting firm broached the subject of cooperative public-private
sustained-yield units early on in its endeavors. The forest industry was so

intertwined with the Forest Service in the western United States that purely private sustained yield would be difficult to conceive. In one pioneering effort, Mason's firm partnered with a Weyerhaeuser company to attempt to create the first cooperative sustained-yield unit. In the mid-1920s the Clearwater Timber Company, a Weyerhaeuser subsidiary based in Lewiston, Idaho, hired Mason to prepare a ninety-page report on the economics, logistics, and general viability of a sustained-yield management plan for its forestlands.[24] Predominantly white pine, these forestlands had not yet been logged significantly despite their proximity to Weyerhaeuser's Lewiston lumber mill. While concerned that the Forest Service would try to wrest control of planning, the consultants also understood that if it did not establish common ground and communication with the national forest it bordered, the Clearwater Timber Company would be economically unable to maintain sustained yield.[25] When the district forester consented to discussions, the excited Stevens remarked, "This is the first time, so far as I know, where a private concern and the Federal Government have agreed to sit down at the same table and work out a completely cooperative working plan, presupposing joint conduct of a sustained yield working circle and all that that involves."[26] When discussions took place, however, the Forest Service representatives proved reluctant to make the concessions the company requested, or to modify existing silvicultural standards.[27]

Whether the Clearwater Timber Company managed to successfully broker this agreement with the Forest Service, and indeed whether it would ever thrive under permanent sustained yield, were not actually of paramount importance to the company. Mason and his colleagues had designed a strategy that quietly hedged the company's bet on sustained yield by overharvesting in the first cutting cycle.[28] A consultant conceded privately that if Forest Service cooperation fell through, "it will still be possible for the company to go out of business at that time and still be money ahead, for the first cycle has been designed to be economically sound in itself." If the Forest Service refused to cooperate, or backed out of an agreement, the company would still exit the sustained-yield agreement with a profit. "It is not, therefore, essential from the standpoint of the company that this operation become perpetual."[29] Consultants did, however, observe that achieving perpetual sustained yield in the area "is particularly desirable from the point of view of the public, and above all of the local communities" that relied heavily on logging for employment. The Forest Service negotiators did sour on the company's demands and declined to enter into any agreement with them. Although its special cooperative agreement failed, the company valued public approval and did not implement its hypothetical exit strategy. Instead, it used standard federal logging contracts to fill out its planned cutting cycles. The Clearwater lands were thus managed for sustained yield, although the company may also have been operating at a loss. Cooperative sustained yield, despite clear benefits to local communities, would be more difficult to grasp than Mason had hoped.[30]

Committing to region-wide sustained yield meant giving up autonomy, which no single company could afford to do unless all of them did. Industrial sustained yield could be achieved not just through the 1929 "lumber Mussolini" plan but also by installing government-mandated external controls on logging activity. Such a plan became more feasible after Roosevelt's election, as the New Deal brought with it a host of new regulatory controls. Article X of the Lumber Code of the National Recovery Administration, passed in 1933, made some provisions for yield control within its larger goal of enforcing good forest practices. However, as discussed in a previous chapter of this book, challenges to the constitutionality of the NRA began as soon as it passed, and most of the Lumber Code forestry mandates were never realized. Greater success would be found in linking sustained yield to the New Deal efforts to transform public land management. Instead of monitoring the harvest, as the NRA would have required, a sustained-yield management plan could focus on monitoring the lands. A district of public forestland, already subject to strict oversight and regulation, could be placed into a program of sustained-yield management. Regulating the harvest could then consist of admitting or barring loggers' entry into specific places on the map, depending on how much logging foresters believed would be optimal for sustained yield. While lumber companies were always in it for the money, many were stable enough that they did not rank maximum immediate profit as their top priority. Further, sustained yield might favor the largest and most stable lumber companies, as small concerns would have difficulty gaining a niche in a market where their activities were tightly regulated. Nonetheless, the Lumber Code as a whole was considered a major impediment for the industry, and its dissolution was hardly mourned.[31]

The first real success with public sustained yield in the United States came not on the federal level but with the Oregon and California Revested Lands Sustained Yield Management Act of 1937. Spread out a map of Oregon and it will reveal curiously rectilinear checkerboards of green and brown scattered through the Coast Range and the foothills of the Cascades, stretching south into the complex topography of the Siskiyous. These squares form the O&C lands, visible traces of the failed railroad land grants of the nineteenth century. The peculiar arrangement of these forestlands originated in the turmoil surrounding the establishment of a rail line from Portland to California, called the Oregon and California Railroad. In 1866 the federal government granted public domain lands to the O&C to create a rail link between Portland and the wealthy boomtowns of California. These lands were divided in the checkerboard pattern common to many railroad land grants of the time, but their sale was restricted in ways that the federal grants for the major transcontinental railroads had not been. Hoping to encourage the settlement of the Oregon country, Congress had stipulated that these checkerboard lands could be sold only to "actual settlers" rather than to lumber companies, land speculators, or other large-scale investors. As the General Land Office was still transferring nearby public lands to Oregon homesteaders for much less than market value, however, the railroad could have little hope of selling its less desirable acreage. Worse still, much of the O&C land was rugged and inaccessible, covered with massive Douglas fir no homesteader could hope to remove, and thus impossible to imagine as valuable for settlement.

Rail laying commenced down the length of the wide Willamette Valley with little challenge, and the line quickly became integral to western Oregon's burgeoning transportation needs. However, as the builders began to contend with the rugged ranges that drew in around the south end of the valley, their momentum foundered. Despite financial and legal turmoil, the line was completed, but most of the land-grant checkerboard was still unclaimed. While much of the checkerboard was thickly covered with Douglas fir, the stipulations precluded the railroad from openly selling to lumber companies. To avoid paying taxes on unsalable land, the railroad postponed taking possession of the checkerboard land for years. Left in legal limbo, precluded from sale or use, the O&C counties were not only left without their anticipated increase in tax income and inhabitants, but sometimes even bore significant costs for maintaining the untended lands and managing access to the publicly and privately owned lands interspersed among the squares. The possible fate of the O&C lands was a perennial source of discussion among the state's foresters, politicians, and businessmen. As the railroad owners were charged with corruption and mismanagement in the early years of the twentieth century, the ownership of the forests remained undecided. It was not until an act of Congress in 1916 that the state could finally break the lands out of railroad ownership. The checkerboard of forestlands was formally transferred to the control of the General Land Office, which reopened the land for sale, but there were still few who wanted the tax burden of these lands when logging leases could be had elsewhere for less. Thus, little money came to the local communities. Decisions for managing revenues and access to this complex forest remained contentious for nearly two decades, as much of the most accessible and valuable timber on the now-public lands was cut.[32]

In 1936 a change was finally made. Secretary of the Interior Harold Ickes was increasingly subject to criticism that his General Land Office was bungling its management of these forests. At the same time, he was determined to engineer the

transfer of the Forest Service and all its lands from the Department of Agriculture
to the Interior, solidifying control of public lands under his administration. Ickes
believed that adopting sweeping and conspicuous change in the approach to
these forests would not only slow the pace of their decline but also prove that
the Department of the Interior was capable of effectively managing forestlands.
The O&C Sustained Yield Act (1937) imposed a formal policy of sustained-yield
forestry on the lands, putting them under coordinated land management. Similar
in many ways to the better-known Taylor Grazing Act (1934), the O&C Act
facilitated coordination of fire management, access, and logging under centralized
control. The goal was to maintain the lumber output as a whole at a natural
replacement level by carefully regulating logging leases on the various sections.
The act was heavily influenced by input from David T. Mason, who had been
discussing sustained yield on the O&C lands for over a decade.[33] He saw sustained
yield on the O&C lands as a first step toward instituting such management much
more widely on federal forestlands. The industry shied away from independently
enacting sustained yield in its activities but willingly took on logging contracts
within lands that hewed to the sustained-yield management plan. In order to
work, sustained yield required tightly controlled management and a complete
commitment to long-term planning, and sustained-yield management had
previously been enacted only on industrially owned lands. The O&C lands thus
became the experimental laboratory for public-land implementation of sustained-
yield forestry.

Weyerhaeuser and the Birth of Lumber Public Relations

In the years since Frederick Weyerhaeuser's turn-of-the-century move into the
Northwest, his conglomerate's power and influence in the region had grown
large, inviting scrutiny of both the companies and the family. In the 1930s,
the Weyerhaeuser companies were not publicly held but rather owned almost
entirely by members of the extended family, meaning that corporate and family
finances were tangled together. As a habit, company executives avoided press
coverage and the company tried to stay out of the public eye as much as possible.
Weyerhaeuser kept its corporate finances so private that only seven copies of its
1932 annual report were produced. Nonetheless, the *Seattle Post-Intelligencer*
still published sets of articles in 1935 and 1936 detailing both Weyerhaeuser
family and corporate wealth. The Weyerhaeuser tradition of avoiding local press
coverage and staying out of the public eye was becoming more difficult as the
company's profile rose. After George, the young son of John Philip Weyerhaeuser

Jr., was kidnapped for ransom in 1935, his father had another reason to hate the press. Then the executive vice president of the Weyerhaeuser Timber Company, he blamed the *Post-Intelligencer* for inspiring the kidnapping through its extensive coverage of both the family's wealth and its Tacoma address. After his son's safe return, he complained that reporters were "one class of animal which never seems to be apprehended. . . . They stooped to *any* device to get something to write about."[34] J. P. "Phil" Weyerhaeuser Jr. made most of the decisions about Weyerhaeuser's day-to-day operations in the Northwest. Soon after these press fiascoes, he decided to fight back by constructing his own corporate publicity rather than continuing the company's tradition of resisting public attention.[35] Phil Weyerhaeuser also suspected that the paper's reportage negatively affected the Weyerhaeuser reputation with both the general public and the company's blue-collar employees. The company's financial information, he worried, "in the hands of our Union Committees will be embarrassing . . . demands of all kinds are being presented to operators, ourselves included."[36]

Instead of allowing their actions to be the subject of news stories over which they had no control, Weyerhaeuser executives began releasing information to the public in order to control the context the stories were given. In contrast to the 1936 *Post-Intelligencer* exposé of Weyerhaeuser Timber Company finances, in 1937 Phil Weyerhaeuser released the annual report to all of its operations managers and shop foremen, accompanied by a ten-year review of the company's finances. He explained this decision to his uncle Frederick E. Weyerhaeuser, writing, "I have the feeling strongly that in one way or another much information about the Company, the amount of its dividends in particular, will be known to the public either through the publication or duplication of the report, such as occurred last year" or through reports to the Securities and Exchange Commission or other offices. "That being so, it seems to me to be prudent to explain the situation in the way outlined, the more so because the story is a good one."[37] This move to publicize Weyerhaeuser finances troubled F. E. Weyerhaeuser, the company president, who worried that his nephew was taking too much advice from public relations men by being so brazenly open. His main concern was whether the company could articulate its position well enough to counteract any negative interpretations of it. While he worried whether any good could come of sharing the company's financial information, he did not stand in the way of his nephew's attempts to improve the Weyerhaeuser name's public appeal.[38]

The first step in Weyerhaeuser's new corporate public relations was centered on the Douglas fir forest. Public relations director Roderic Olzendam had been hired in 1937 and became the leader of the company's audacious embrace of the new arts of public relations. Olzendam's job title was labor relations counselor, although his main duties were, from the beginning, to develop public relations strategies for the Weyerhaeuser companies. He rationalized his job title by saying his duties at Weyerhaeuser were "to provide steady employment to the largest possible number of persons under the best conditions."[39] Weyerhaeuser's new drive to control public opinion resulted from the scrutiny and pressure the company faced from foresters and politicians. An increasingly pointed critique of the lumber industry's methods from foresters and other conservationists had created enough public concern that a response was necessary. Works like George P. Ahern's *Deforested America* had marshaled statistics to create a fuller picture of the nationwide scope of lumber industry activity. Authors like Bob Marshall and Gifford Pinchot had used their prominence as foresters to bring attention to their calls for increased regulation of industrial forest practices. The past decades had seen the states and federal government increasingly monitoring and regulating aspects of the timber industry. If the general public became too disenchanted with the timber industry, further curtailment of its activities could be forthcoming. Severe restrictions could be disastrous for the industry, so attempts at controlling the public image of the Weyerhaeuser corporation through the newly defined field of public relations could be a necessary hedge against future problems.[40]

These years were the first in which large companies imagined "public relations" as a necessary endeavor for corporate success. The Great Depression and the policies of the Roosevelt administration had brought corporate behavior under closer scrutiny than it had ever been before. Hoover's policy of allowing corporations to operate with little governmental oversight was blamed for fostering the unstable conditions that led to the Depression; hence many of the Roosevelt administration New Deal programs were concerned specifically with regulating commercial enterprises. Furthermore, a number of industries, including the lumber industry, continued to have large-scale, widely reported troubles with labor unions. The result was an increasing tendency among both politicians and the general public to view industries and large corporations as enemies of the American way of life. Governmental initiatives such as antitrust laws and graduated income taxes all had one common underlying sentiment: the behavior of large corporations, and the wealthy businessmen who controlled them, were contrary to the prosperity of the American people as a whole. President Roosevelt assailed large American corporations for not conducting their business in the name of the public good, stating before Congress that industries were "a concentration of public power . . . masking itself as a system of free enterprise."[41] In response to these currents of public opinion, a new sort of business practice, the corporate public relations campaign, came to prominence in the 1930s.[42]

Weyerhaeuser's initial "Timber Is a Crop!" advertisement. Seattle-Post Intelligencer, July 25, 1937.

The corporate public relations campaigns of the mid to late 1930s were intended to show the average American that big business, and the capitalist economic framework as a whole, were not the scapegoats that New Deal rhetoric made them out to be. Indeed, many industrialists and political conservatives saw such campaigns as simply a response in kind to the public relations campaigns of the Roosevelt administration. In order to better convince the American people of the necessity of the New Deal, Roosevelt employed media in ways previous administrations had never conceived. His methods ranged from the *Fireside Chats*, broadcast weekly on radio, to documentary short films such as the Resettlement Administration–funded *The Plow That Broke the Plains* and the Tennessee Valley Authority–funded *The River*, to the Farm Security Administration–funded photographic essays of Dorothea Lange. The Roosevelt administration employed Edward L. Bernays, considered one of the greatest public relations experts of the twentieth century, in some of its New Deal enterprises. In response, industrial leaders began to rethink the ways in which they conceived of publicity and advertising. While previously corporations would advertise their individual products directly to consumers, there was a growing sense that advertising and other methods of public relations could be used to sway public opinion on a larger scale. Beginning in 1934, the National Association of Manufacturers, employing a number of public relations specialists, began a new effort to forge industrial policy and public relations. The most prominent of these efforts was a campaign entitled "The American Way," which aimed to show the average American the value and power of the capitalist system. Its advertisements reframed the rhetoric and ideals of the New Deal in service of industry, as in a 1936 advertisement proclaiming, "Our American plan of living is simple. Its ideal—that works—is the greatest good for the greatest number. . . . Our American plan of living is pleasant. Our American plan of living is the world's envy. No nation, or group lives as well as we do."[43]

Olzendam delivered a speech entitled "Timber Is a Crop" before the 1937 meeting of the Pacific Logging Congress, an annual trade meeting of Pacific Coast lumber industry executives. The Weyerhaeuser Timber Company reprinted this speech later that year in an extremely lavish pamphlet. Olzendam's talk was intended for both the assembled participants and the inhabitants of the region. In it he presented the Weyerhaeuser philosophy of reforestation and addressed the larger problems faced by the lumber industry regarding public opinion. In this talk Olzendam endeavored to answer the question he posed to his audience: "What is the combined Voice of Public Opinion saying about the forest industry, about you and me and all our thousands of associates in it?"[44] In answering this question, he laid out the company's plans for massive reforestation projects in pursuit of a permanent forest industry in the Douglas fir region and around the nation. He presented these answers not simply as one company's answer but instead as "a first, rough draft of the answer of the forest industry to this indictment by the Voice of Public Opinion. We should try to reach a point where the forest industry and public opinion can meet on common ground and talk in friendly terms."[45] In this way, Olzendam, and through him the Weyerhaeuser Timber Company, were articulating an industrial forest policy more clearly than had been done previously. Through putting the answers into words, Olzendam began to shift the message from a Weyerhaeuser-specific slogan to a message that could apply to the entire forest industry.

Olzendam's answers, given in the guise of, as he put it, "The Voice of the Forest Industry," were meant to give the assembled lumbermen an arsenal of potential retorts to any criticism that might be leveled at them. He defended the forest worker by saying, "Not a man in the forest industry has a grudge against a tree. . . . He isn't vicious in his work. He isn't a deliberate despoiler of beauty. . . . To condemn the hundreds of thousands of people whose life paths are in the forest industry is unthinking and unjust. To go ahead and freely use the thousand and one products of the forest on top of this denunciation is really not square."[46] Handling timber as a crop, Olzendam said, meant that the logged-off lands would regenerate, evoking "this new, vigorous, green crop heading for another harvest *80–100 years* hence—a long cycle, but a sure one, barring fires and pests."[47] He stressed the importance of cooperation between landowners and concessions from the public, the state governments, and the federal government. "We believe we have caught the situation in time. There need be no future period of timber famine between the cutting of the last big trees and the harvesting of the first new crop, if there is close and sympathetic cooperation between private timber owners, governmental agencies, and the general public."[48] Olzendam addressed both the fears of the public, who were more likely to see the devastation of the clear-cuts than to notice any seedlings sprouting up, and the worries of the lumber worker, who foresaw a day when the timber jobs would all be gone. His speech concluded with an evocation of the dependence of so many in the Pacific Northwest, including his listeners, on the permanence of the region's forest industry. "Timber is a crop," he repeated once again. "Those four words carry more meaning for the men and the women who do the work of the forest industry than any other four words in the language. . . . The policy of handling timber as a crop will do more in the long run to fulfill the desires of the average working man in the forest industry than any other one thing."[49]

A new slogan for the Weyerhaeuser companies, "Timber Is a Crop!" was introduced to the public on July 25, 1937, in a full-page advertisement in the *Seattle Post-Intelligencer*. The appearance of the advertisement, as well as its content, was Olzendam's doing. While Olzendam publicly took credit for the creation of the company catchphrase "timber is a crop," that phrase and the idea behind it had both been in circulation in regional lumber industry discourse. Historian Charles Twining writes that Phil Weyerhaeuser himself had suggested that phrase to Olzendam soon before the publication of the *Post-Intelligencer* advertisement. However, Phil Weyerhaeuser was not the originator of the phrase either, as the first public use of it by Weyerhaeuser personnel was in 1909. George S. Long, the first general manager of the company, had used it that year in a speech at the annual meeting of the Pacific Coast Lumber Manufacturers' Association.[50] And indeed, others had also often used variations of the phrase in relation to the future of the American lumber industry. William B. Greeley, head of the Forest Service from 1920 to 1928 and a vocal proponent of industrial forestry, in 1922 envisioned "shifting our source of timber from the supplies stored up in virgin forests which are sought and mined out in order of their accessibility . . . to successive timber

crops grown in the 39 States which contain large areas of forest land."[51] Thornton
T. Munger also saw the possibility that Douglas fir timber could be grown as a crop
in the Pacific Northwest. Commenting on industrial reforestation experiments in
the 1920s, he wrote that "this is the kind of knowledge foresters must have to put
forestry on as scientific a basis as is pineapple growing or loganberry culture."[52]
The Capper Report, a Senate-commissioned study of the nation's forests, also
made use of agricultural language in its exploration of the nation's prospects
for future forest availability. A widely distributed 100-page article summarizing
the Capper Report's findings, coauthored by seven Forest Service foresters, was
titled "Timber: Mine or Crop?" In light of this title, the Weyerhaeuser motto
could be seen not as an innovative new concept but as an answer to a question
that the Forest Service had posed a decade and a half earlier.[53] Indeed, foresters
had often used agricultural language to describe forest management, especially in
public forums.[54] The Weyerhaeuser declaration of the statement was not a new
concept for any person who followed news of American forests. Gifford Pinchot
had used agricultural language to describe the American forest during his tenure
as chief of the Forest Service. In 1901 he had laid out his principles of forestry,
the first two of which clearly approached timber as an agricultural commodity:
"First: The forest is treated as a working capital whose purpose is to produce
successive crops. Second: With that purpose in view, a working plan is prepared
and followed in harvesting the forest crops."[55]

The adoption of "Timber Is a Crop!" as the company's advertising slogan
heralded not the culmination of Weyerhaeuser's mission to determine its perception
in the eyes of the general public, but the beginning of a new campaign toward that
goal. Company executives' satisfaction with the response to the campaign led them
to authorize additional messaging in the same vein. Instead of advertisements that
focused on the quality of Weyerhaeuser wood and paper products, new publicity
described the long-range future of the company and the forest. Conventional print
advertising might affect sales but did little to change the public's opinion of either
the company name or the lumber industry as a whole.[56] In a move that seemed
directly inspired by the Roosevelt administration's half-hour documentary films of
1936 and 1937, the Weyerhaeuser Timber Company produced its own half-hour
film, *Trees and Homes*, in 1938. *Trees and Homes* was an entertaining, expensively
made examination of the importance of wood and wood products in American
life, apparently exhibited mostly in educational settings.[57]

Pacific Northwest conceptions of the Douglas fir forest began to shift from
viewing timber as a forest resource to viewing the forest as a nearly monocultural
cropland. If the site of production was a farm, then the product itself was obviously
a crop. Claiming "Timber Is a Crop" served to further the commodification

of the Douglas fir and other such high-yield timber trees. The metaphorical transformation of the Douglas fir into a crop implied a certain reliability and consistency in the supply of timber. While agricultural crops might suffer annual fluctuations in productivity due to the vagaries of weather, longer-term projections for agricultural crops were more consistently positive. This contrasted with the European and early American concept of the timber famine, an alarmist fear that the long-term projections for timber production were a complete collapse of the industry. By reframing timber as a crop, the timber industry could reestablish public faith in the industry's long-term stability. Furthermore, if timber was a crop, then the site could be seen solely as a location for timber production rather than as a multiuse, multispecies ecosystem. The forest was increasingly being managed by state and federal government regulations, and forests held a particular resonance among the general public. By describing their lands as farms, the polar opposite of wilderness, those in the timber industry could perhaps fend off further regulation of their practices.

By invoking agriculture to describe reforestation, Weyerhaeuser aligned its replanted lands with the midwestern and eastern sense of farmland rather than with the western sense of the open frontier. Well into the twentieth century, much of the Northwest's forestland was still untended and open, nominally parceled out to obscure absentee owners or held by large government agencies like the General Land Office. Even in the 1930s, many Pacific Northwesterners still thought of their local forests as essentially public, with game, berries, and even timber resources free for the taking by anyone with the desire to remove them.[58] A farm was private land, not to be entered without permission, and the unauthorized removal of crops or other resources from farmland was illegal and immoral. Olzendam acknowledged this:

> The moment I say "farm" there immediately flashes in the mind of every person . . . a mental picture of his own personal idea of a farm. A farm is an area where man grows successive crops of corn, wheat, hay, oats, and vegetables. . . . All of us have a high regard for "farmers." We respect the farmer and we respect his farm. We do not wander idly through his wheat fields, nor do we picnic in his strawberry patch. It would never enter our minds to wander in and pick his grapes without his permission. . . . Well, Timber is a Crop, just like any other crop, except that it takes a longer time to grow a crop of trees than it does to grow a crop of potatoes.[59]

Such statements were meant to help keep the public from interfering with the activities on the Tree Farm, and at the same time increase the public respect for the enterprise. The ideal of eternally productive forestlands also defined these lands as sharply different from the ideal of the open, untouched forest held by wilderness activists and park builders. The wilderness rhetoric of the mid-1930s reaffirmed the sensibility of earlier preservationists: unsettled forestlands were uncommercialized places where individuals were free to roam. If a forest could be a crop, however, then not all forestlands would necessarily be wild. When foresters called reforestation a kind of agriculture, it drove home the fact that these reforested lands, while they might return to being forest, would not return to being wilderness.[60]

Challenging Times for Industrial Forestry

The debacle of the NRA changed the nature of the Roosevelt administration's relationship with the nation's industrialists. The administration tried new methods of stabilizing American industries and enforcing good business practices. Without the need to cooperate with the goals of the NRA, relations between industry and government became more confrontational. In 1936, Congress passed the Robinson-Patman Act, legislation that strengthened the controls on price fixing in the Clayton Antitrust Act. This set the stage for investigations of industrial lumber practices by the Federal Trade Commission, the Department of Justice, and other federal entities.

The suspicion of price fixing in timber sales was based on the existence of a great deal of communication between the major US lumber companies. These avenues of communication had been developed in response to the instability of lumber prices and then strengthened in response to the Lumber Code of the NRA, when cooperation on prices had been mandated by the federal government. Although in 1935 the entire NRA had been declared unconstitutional by the Supreme Court, the lumber industry was still suffering from production decreases of 44 percent since 1929. However, some of the conservation clauses of the code would have benefited large companies by limiting oversupply, which could have inflated retail prices for lumber. In addition, some of the wholesaling and labor provisions would have put additional pressure on the smallest operators, possibly driving them out of business. The lumber giants voluntarily decided to retain certain aspects of the NRA's Lumber Code in an attempt to stabilize the price of lumber. Accused of price fixing and attempting to reduce competition in the industry, the Weyerhaeuser companies were the subject of an investigation by the Federal Trade Commission in December 1937. The investigation yielded no charges, however, and was terminated the next year.

At the same time that industry giants' lumber pricing was being scrutinized, federal inquiries also aimed to determine whether government regulation of the logging of large lumber companies on their own private lands should be increased. A Joint Congressional Committee on Forestry was convened at President Roosevelt's request in June 1938 and began a series of public hearings to investigate forest practices around the nation. Mindful of its prominence in the industry, the Weyerhaeuser Timber Company carefully orchestrated its participation in the committee hearings. Company forester C. S. Chapman wrote Phil Weyerhaeuser with a plan to dominate the hearings with the Weyerhaeuser point of view in every region nationwide, suggesting that "in each region we can call on certain individuals to develop special phases of a program. If . . . anyone who desires goes before the committee, things we don't desire are apt to be brought out and a large and confusing record result."[61] Phil Weyerhaeuser also suggested bringing the public relations committee of General Timber Service, a consulting and promotional agency allied with Weyerhaeuser Sales Company, into the discussions to help craft their message.[62] At the Pacific Northwest regional hearings, the chairmen of all the major Weyerhaeuser subsidiaries, including Boise Payette, Potlatch Forests, Northwest Paper Company, and Weyerhaeuser Timber Company, testified before the congressional committee. The message from the leaders of all the subsidiary companies was the same: increasing government regulation would be detrimental to the region's economic vitality. When its investigations drew to a close two years later, the committee recommended voluntary cooperation rather than outright regulation. However, the committee report also observed that it was the people's prerogative to impose more stringent changes through legal means if desired.[63]

In late 1938, the Department of Justice declared war on "the blighting encroachment of monopoly," as Assistant Attorney General Thurman Arnold declared his intention to "de-NRA-ize" American industries.[64] The Department of Justice amassed documents from a large number of lumber industry entities, which together suggested a pattern of price fixing. Under the aegis of the Sherman Antitrust Act, the Justice Department indicted 6 lumbermen's trade associations, 167 logging and lumber corporations, and 53 individuals on production curtailment and price fixing. Among those indicted were 7 of the largest and best-known Weyerhaeuser subsidiaries in the Pacific Northwest's pine and Douglas fir regions. Rather than face criminal proceedings, the Weyerhaeuser companies, along with many others, signed a "consent decree," in effect an admission of wrongdoing and a pledge to change their practices. While financially prudent in the short term, the long-term effect of this admission was to deter any future attempt at a collective effort toward sustained yield on industrial lands.[65] First the NRA as a whole had been declared unconstitutional infringement on the rights of the free market, and now the independent adoption of these NRA provisions was declared a violation of trade laws as well. The government's decision that a cooperative industrial sustained-yield management agreement constituted a form of monopoly chilled interest in pursuing that avenue. The federal scrutiny demonstrated the lucrative potential of unregulated cooperative sustained-yield arrangements but also proved that a different approach was needed to avoid running afoul of regulators. These years prompted industrial foresters to reappraise their conceptions of a private cartel of sustained-yield operators in the Douglas fir.

Challenging Times for Radical Forestry

Amid all the turmoil of these years, one group of critics that had been loud earlier in the decade became quiet. The Zon coterie's attacks on the profession's priorities, so strident before, were barely noticeable in the late 1930s dialogue about industrial forestry. By the late 1930s, the focus of their efforts had evolved from trying to change professional forestry from the inside to seeking new alternatives to that increasingly conservative profession. The habitually argumentative Zon, whose career had arguably been the most affected by his radical politics and outspoken criticism, was now reluctant to confront foresters. Zon was still serving as both the Lake States Forest Experiment Station director and the overseer of the troubled Great Plains Shelterbelt Project from his home in Minnesota. In 1933, he reflected that "I have been thinking a lot about my relation to forestry and foresters . . . I need a new orientation and possibly my past tactics have not been the best." Several months later he confided that he had "been suffering from some nervous breakdown" affecting his digestion, and that "every little excitement is apt to throw me out of balance."[66] In 1935 he wrote to Ahern that he was no longer interested in disrupting the Society of American Foresters, as "it seems to me that the game is hardly worth the candle, considering that the Society really plays such a small part in the present struggle . . . [and] the *Journal of Forestry* is not the only vehicle through which we can speak."[67] Ahern was in poor health,[68] and Pinchot had lost his interest in the fight as well. Zon complained that Pinchot had "decided to take life easy and not to exert himself"[69] after the Zon Petition, but in fact Pinchot's attention had simply been diverted to other causes and campaigns. He was included in less of the correspondence circulating between the members of the coterie and wrote infrequently to them himself. His 1938 campaign for a second term as governor of Pennsylvania demanded much of his time and attention. Following his defeat in that campaign, Pinchot mostly retired from public life and in his last years spent almost no time among members of the profession he had helped create.[70]

Marshall, traveling constantly for his job with Indian Affairs, found that organizing and developing a framework for the Wilderness Society dominated his attention. The Wilderness Society's agenda was the preservation and regulation of pristine natural areas for recreation and research, a goal that did not overlap easily with his earlier aims, as spelled out in *The People's Forests*, of developing intact forests for human use through an enlightened new form of forestry. Marshall continued to try to transform foresters into wilderness advocates, instead of abandoning the framework of forestry entirely. Leopold, for one, thought this was pointless. He admonished Marshall that "if you appointed a bunch of engineers and contractors to be the board of trustess [sic] for an art museum, you would have a situation analogous to the average group of administrative foresters or park officials deciding what to do about a wilderness policy. They mean the best in the world, but they are simply not conversant with the subject."[71] Marshall was undaunted, though, and even proposed allowing some commercial logging within forests. He disputed the notion that "one can still preserve a perfect wilderness or that there ever was a perfect wilderness since the first man evolved,"[72] as if it were merely splitting hairs. "There has never been a completely unmodified wilderness since human beings existed," he wrote to Lincoln Ellison, a forester at the Northern Rocky Mountain Forest Experiment Station, in 1935. "If you want to get tangled up in logic you could sooner or later make it appear that Broadway, New York was essentially as much of a wilderness as the South Fork of the Flathead. Nevertheless, I do not think that's necessary. To conclude: we have to make some practical assumptions and definitions and then go ahead."[73] With his eye now fixed on saving wild forests from the looming threat of clear-cutting, Marshall had little patience for the finer points of regulation that had once consumed him. However, he continued to snipe at the Society of American Foresters in person and in print, even as he acknowledged less enthusiasm for the fight. "What is wrong with us that we permitted such a reactionary slate to be elected to guide the affairs of the Society for the next two years?" Marshall lamented to Zon in 1937. "The trouble is that those of us who belong to the socially minded element of the Society are too damn lazy."[74] Indeed, without Marshall's support and energy, the coterie was loath to indulge in argumentative episodes with powerful foes.

The Zon coterie's final disintegration came amid tragedy. The decade ended with a stunning double blow to radical forestry, as the sudden death of Bob Marshall on November 11, 1939, was followed on December 20, 1939, by the death of another prominent Zon Petition signatory, Forest Service chief Ferdinand Silcox. Both deaths were shocks to all; Marshall was a vibrant thirty-eight when he passed away on a train of an apparent heart attack, and Silcox, at fifty-seven, was at the pinnacle of the profession when he too died of a heart attack. The loss of Marshall meant the loss of the ideological heart of the group, and the loss of Silcox meant the loss of its highest-ranked voice within the Forest Service. Marshall's money and enthusiasm had propelled the coterie's ideals forward in the face of repeated obstacles to success. Silcox's position of actual power as chief of the Forest Service had provided the coterie's recommendations with weight they might not otherwise have had. The remnants of the group reacted to these sudden losses with grief and confusion. The survivors found that their common causes and concerns were harder to keep at the forefront of discourse within professional forestry. The Wilderness Society had depended on Marshall's enthusiasm and energy, as well as his money. While his will provided for the society's financial well-being, maintaining leadership and direction was more difficult. Robert Sterling Yard, who had served as the organizational hub of the society, reluctantly took the helm. Yard was not a forester by profession but rather a writer who had spent a career extolling public lands recreation. Marshall's habit of antagonizing the Society of American Foresters, which had already worn thin for many in the coterie, died when he did. At a time of uncertainty, stabilizing the society would require building bridges, not burning them.[75]

Conclusion

Despite their embrace of industrial forestry and efforts like the "Timber Is a Crop!" campaign, by the end of the 1930s lumber companies still had more work to do to sway national public opinion their way. The depth of mistrust can be seen in a 1941 survey of attitudes toward the lumber industry, funded by American Forest Products Industries and conducted by Opinion Research. This study showed that 62 percent of those polled believed that "forests are being cut faster than they are being replaced," while 52 percent thought that the supply of standing timber was being reduced at "an alarming rate." The majority of these people had no connection to the lumber industry and lived nowhere near lumber towns. Their perception of the supply of American lumber and of American logging practices was based on their impressions of the industry as it had been portrayed by politicians and activists and by press coverage of books like *The People's Forests* and *Deforested America*. Many of those polled piled all the blame for the dire situation of the forests at the feet of the lumber industry, rather than with the Forest Service or the marketplace. Seventy-two percent were certain that "wasteful cuttings" were being made during logging, and most of those believed that the lumber industry was responsible for this waste. Forty-six percent blamed the lumber industry for wasteful practices, and 43 percent believed that the lumber companies were doing little to preserve future forest production through the practice of forestry. Perhaps most surprising, 38 percent of the Americans polled had "never heard of any program to replace drain on the forests" through reforestation, despite the existence of many such programs in federal, state, and private entities. Of the 62 percent who had heard of reforestation programs, 96 percent believed that such programs were undertaken only by the government, not by private foresters.[76] The industry's efforts to publicize its own work had not yet borne much fruit. In 1941, Americans still perceived the American lumber industry as it had been fifty years earlier: ravenous, wasteful, and indifferent to the long-term health of forests.

Deterred from pursuing private sustained-yield management yet also threatened with regulation if traditional clear-cut logging continued, lumber executives felt they were in a bind. Attempts at controlling public perception held promise, but something more tangible than slogans was needed. They needed a new idea that would satisfy both forest management and public relations. While for decades there had been scattered efforts to improve logging practices and long-term stability in response to public pressure, these projects had often received little attention outside forestry circles. Even the *Journal of Forestry* observed that "if one were looking for the best examples of good forest practice in the United States, one would probably find that these examples would be found on privately owned land quite as often as on publicly owned land.. . . Unfortunately, many of these companies have received only local recognition for the progressive steps they have taken to manage their holdings on sound forestry principles."[77] Finding more effective ways to fight negative public perception, while also practicing sound forestry, would become a major goal of the Northwest lumber industry as the 1940s began.

6 Divergent Paths
Wild and Tame Forests in the 1940s

By the late 1930s, the new agricultural rhetoric of industrial forestry had begun to play out on the ground, in actual changes to the management of Douglas fir forests. Olzendam's claim echoed around the region, a counter to federal and state pressure for improvements in industrial forestry. The O&C Sustained Yield Act hinted at the beginning of a new era in public forestry, in which forest use would be managed more restrictively. Industrialists feared that act was the harbinger of an increasingly complex and controlled system of forest management, where logging companies would submit to the restrictions set by long-range planners sitting far from the forest in Washington, DC, offices. Rather than submit to this, lumber companies developed a new type of forest management, a manifestation of their "Timber Is a Crop" cry. The Tree Farm used a set of simplified forestry principles to allow lumbermen to exert heavy control over what trees grew on their land, and how. The Tree Farm idea turned out to be technical enough to appease many observers' fears of forest devastation, while still appealing to lumber executives. Professional foresters looked on with dismay as the superficial, logging-centered Tree Farm idea caught on in the Douglas fir region.

This chapter examines how Weyerhaeuser executives developed the Tree Farm as an idea and brought it to life. On its clear-cut private lands in the Northwest, the Tree Farms were a new kind of landscape—not forest, not farm, and not wasteland. The chapter begins with a close examination of Weyerhaeuser's establishment of the first Tree Farm, and the reimagining of forest management that was necessary to transform an old clear-cut into a new type of landscape. After the first Tree Farm was established, lumber industry executives constructed a network of semiautonomous agencies to manufacture public trust in the Tree Farm concept.

Both academic foresters and state government officials received tree farming with great suspicion. Despite professional resistance to the concept, tree farming spread quickly. Tree Farms not only changed the landscape, they also changed the tenor of American professional forestry. Lumber companies asked the industrial foresters in their employ to do something different from what American foresters had done before: to shift toward a focus on high-intensity silviculture and fire suppression. Whether or not these techniques were well regarded by the larger profession of forestry mattered little to lumber companies. The specialty of industrial forestry grew and changed to accommodate industry's needs. In so doing, the profession oriented itself increasingly toward a focus on lumber production. The wartime demand for lumber exacerbated this focus on production among federal foresters as well.

As the profession splintered and the radicals were edged out of power, the Douglas fir forests were left to be managed as a commodity rather than a natural landscape. While the postwar history of sustained-yield management on public lands has been examined often, the slightly earlier development of this alternative form of forest management on industrial lands has been mostly neglected. This chapter approaches the Tree Farm as the industry's response to mounting pressure from academic and federal foresters to use forest resources wisely. By 1944, the argumentative, experimental era of professional forestry had been left behind. The federal government, like the lumber industry, moved toward a commodified view of the forest during these years, culminating in the passage of the Sustained Yield Forest Management Act. Those who still called themselves foresters were now in a profession increasingly devoted to lumber production at the expense of more holistic forest management. The profession's organized internal opposition of previous years, originating with people like Zon and Marshall, had dissipated as many of those people found other subjects of focus. The expansion of ecological specialties lured some of those detractors to academia, while the strengthening recreational focus of wilderness activism drew radicals away too. The success of the Tree Farm idea paved the way for the reductionism and commercialization of forest management seen in postwar professional forestry. The fight for these forests ended as their management became simply a quest to grow and harvest the money tree.[1]

Tree Farm Number One: The Clemons Tree Farm

The modern industrial Tree Farm, as a privately funded method for second-growth lumber production, has its origins in the Douglas fir forests of the 1940s. Spearheaded by Roderic Olzendam, Weyerhaeuser's first dedicated public relations executive, the Clemons Tree Farm was designed as a public showpiece as much as a site of lumber production. This first Tree Farm was a manifestation of the Olzendam-designed 1937 advertising campaign's assertion that "Timber Is a Crop!" The Tree Farms would allow the region's largest lumber companies to mold public opinion of company activities as well as control locals' interactions with company-owned land.[2] The Clemons Tree Farm was the first direct planting to be widely known in the United States by the name "Tree Farm." It was situated southeast of Aberdeen, Washington, in the low, rolling mountains of the Coast Range in Grays Harbor County near the town of Montesano. Charles

F. E. Weyerhaeuser, youngest son of the founder, examining a shipment of lumber in a publicity photograph (1930s). Beginning in 1934, as president of the Weyerhaeuser Timber Company, he oversaw decisions on sales and marketing as well as on tree farming and land use. Courtesy Minnesota Historical Society.

Clemons began logging Douglas fir on the site in 1901. The Weyerhaeuser Timber Company purchased the Clemons Logging Company in 1919 and reorganized it as a subsidiary while retaining its original name. Lumbermen and foresters commonly believed that this area of the Coast Range produced the highest volume of wood per acre in the Douglas fir region. Furthermore, the logs and lumber made from Douglas fir trees in this area were thought to be generally of higher quality than those of other areas in the Douglas fir region.[3] Observers from the lumber industry believed that natural reforestation would restock the logged area. The first few years of timber harvesting had been small in scale, and it appeared that Douglas fir seedlings were growing on the clear-cut sites without intervention from foresters. However, the rate of logging increased after Weyerhaeuser acquired the lands, and improved logging techniques and equipment encouraged more and larger clear-cuts on the Clemons site. Several fires swept through the debris-strewn logged-off areas of the site in the early 1930s, destroying all the young regrowth of Douglas fir and other forest species. The combined effects of these changes on the Clemons lands arrested whatever natural reforestation might have been occurring before the Weyerhaeuser acquisition. No effective reforestation was documented on these lands in the years following the fires, and consequently the soils and vegetation declined drastically in quality. In 1936, with most of the old-growth Douglas fir on the land logged off, the Clemons Logging Company's minority stockholders were bought out, and the company was absorbed into the Weyerhaeuser Timber Company.[4]

The Clemons Tree Farm was created through arrangements between the Weyerhaeuser Timber Company, based in Tacoma, Washington, and the US Forest Service. Within the boundaries of the Tree Farm, Weyerhaeuser owned 70,000 acres in 1941. The rest of the land had a patchwork of ownership, with approximately 15,000 acres of state-owned land, 15,000 acres of county-owned land, and 30,000 acres held by miscellaneous other public and private entities. To make the Clemons Tree Farm a contiguous area, Weyerhaeuser obtained leases for the lands that it did not own outright, allowing the entire area to be under the same management regime. Keeping some of the land in public ownership gave the Tree Farm a character of access and cooperation even though the entire area was under the company's control. Altogether, the contiguous land of the Tree Farm covered 130,000 acres, almost all of which had been covered with Douglas fir forest in 1900. Although in a 1938 report a Forest Service surveyor had concluded that there was no hope of reforestation on the Clemons site, the Weyerhaeuser Timber Company went ahead with its plans to make it the nation's first Tree Farm.[5] The acreages owned by both Weyerhaeuser and the public at the site had been almost entirely logged off by the late 1930s using high-lead clear-cutting techniques. Additionally, in 1937 Weyerhaeuser Timber purchased adjacent logged-off land owned by three other unaffiliated lumber companies to increase its holdings in the area. In October 1938, the Forest Service surveyed the status of forest cover and reforestation efforts on a number of sites owned by subsidiaries of Weyerhaeuser, including the Clemons tracts. Part of an ongoing effort to monitor the state of both public and privately owned forestland, this survey found many shortcomings at Clemons. The government surveyor wrote that despite the fact that in 1938 Weyerhaeuser had "spent several thousands of dollars in disposing of accumulated slash for the purpose of reducing risk and protection costs . . . [its] forest practices [were still] not satisfactory chiefly because of hazardous accumulations of unburned slash after logging." The surveyor concluded that "recurrent fires have left many older cutover areas in non-productive condition. It is believed that the operation will be cut out in another two or three years. No possibility for sustained yield in this operation because of lack of mature timber."[6]

Weyerhaeuser designed the Clemons Tree Farm not just as a site of production but also as a place for the company's foresters to conduct Douglas fir research and experiment with different reforestation protocols. While they planted most of the acreage according to their most trusted methods, company foresters also used some of the land to conduct long-duration trials of different silvicultural treatments and planting patterns. In other Weyerhaeuser landholdings, there had been tension between the logging crews and company foresters attempting to begin reforestation projects. Foresters' requirements for reforestation often ran counter to accepted logging practices, prompting disagreements and even labor problems on the logging site. Since all logging operations on the Clemons lands were completed and the land had been designated as unproductive and essentially worthless, company foresters could be given free rein to experiment. Forestry research in the first years of the Clemons Tree Farm centered on perfecting artificial reforestation techniques suitable for different kinds of terrain. That research was accompanied by inquiries into soil fertilization, disease prevention, and pruning and thinning Douglas fir on the Clemons Tree Farm. Some small-scale experiments in breeding genetically superior strains of Douglas fir were also conducted.[7]

What most set the Clemons apart from earlier industrial forestry was the close monitoring and long-term accountability of the trials. The company foresters saw the lands of the Clemons as the perfect place for "a centralized research program to supervise and coordinate the collection of necessary data and information, and to sift out, supplement if needed, and make available with necessary recommendations, the pertinent information gathered by private and public agencies engaged in forestry and forest-products research."[8] The Clemons became a site with two long-term goals: the industrial-scale production of timber and the rigorous study of reforestation. While some mature Douglas fir were spared for natural reseeding trials, most of the foresters' efforts involved labor-intensive artificial reforestation trials. Teams of laborers would plant Douglas fir seedlings to be grown in the company's nursery at Longview, Washington, and experiments in rodent control and thinning of seedlings were planned for the young trees as they grew. Weyerhaeuser foresters closely monitored the activities on the Clemons acreage and planned experimental trials to last for decades. The stated intention was to make the Clemons planting trials the most rigorous industrial forestry experiments to date.[9]

Weyerhaeuser emphasized fire suppression for its Tree Farm program from the beginning. The Copeland Report had published generally accepted estimates of the amount of land that could burn each year and still ensure natural reproduction. For Douglas fir forests, the highest percentage of land that could burn in an average year was 0.3 percent. However, Weyerhaeuser's own estimates of the effect of fire on the land showed that the area being burned annually was about 2 percent.[10] If the Copeland Report's fire statistics were correct, then fire protection would be the most important part of making the Clemons Tree Farm a long-term success. In 1938 and 1940, two large fires swept through the logged-off areas of the Clemons lands, reducing the chances of natural reforestation on yet more of its acreage. Natural reforestation, although very inexpensive, was slow and unpredictable, and the degradation of the site meant that banking on natural regrowth was too risky. The Weyerhaeuser Reforestation and Land Department conducted a study of the economics of fighting forest fires in young-growth forest. They calculated that the 1940 fire cost Weyerhaeuser $15,000, since the labor costs of containing and extinguishing the fire by traditional methods were $6,000, and the labor and materials costs for replacing the young trees destroyed by the fire were approximately $9,000. That amount of money, the company's agents pointed out, could purchase five state-of-the-art fire trucks. If the company had invested in such trucks, as well as in fire prevention campaigns and fire lookouts, fires on the entire Clemons area could have been almost completely averted. Weyerhaeuser's Reforestation and Land Department designed a plan for fighting fires on Weyerhaeuser land that went far beyond the norm of wildland firefighting. Fire crews, trained in working in rough forest terrain, would be on call day and night, waiting for fire alarms relayed by radio. Narrow dirt roads crisscrossed the landscape, ready to bring tanker trucks within reach of any fire within ten minutes of its report.[11] An internal Weyerhaeuser Timber Company report later explained that the Clemons "plan was based on fully stocked land. Unless we had fully stocked land to protect, such an expensive system was not feasible; therefore, plans for restocking, by planting all denuded area was included."[12]

Tree planters and their rows upon rows of tiny Douglas fir made for captivating images, but the barely visible activity of fire suppression was just as important to the process. This complex, expensive, and important work of fire suppression moved the Tree Farm even further away from natural forest. Wildland fire, of course, had long been present in the Douglas fir region, caused by both lightning and human activity. During the dry summer months, patches, sometimes of many acres, might burn in any year. A low-intensity ground fire, the most common type, might cause little long-term damage in a healthy, mature Douglas fir forest. Only a rare, hot crown fire, intensified by wind, would be likely to destroy the largest trees. Hot, fast, destructive fires occurred more frequently on recently logged areas, where there was little moisture accumulated in the soils or plants to control the intensity of the burn. A hot, high-intensity ground fire on such a site could destroy the seedlings and the ecologically important duff as well as the remaining mature trees. In an area recently planted with seedlings, a fire would wipe out not just the vegetation but a large investment of money and labor as well. Along with innovative wildland firefighting techniques and equipment, fire control in the Douglas fir region depended on the cultivation of public cooperation and conscientiousness. Because fires could spread so rapidly in the dry summers typical of the Douglas fir region, it was necessary to keep fire out of not just one's own land but that of all one's neighbors as well. Weyerhaeuser aimed for the fire protection in the Clemons to approach the protection ensured by a city fire department. He told the governor that the Clemons Tree Farm would show that "the same type of prevention and protection system of fire control practiced in cities can be carried out equally well in our forest growing areas."[13] One commentator wrote that "the Tree Farm movement was accomplishing a feat of indoctrination unparalleled in conservation history. . . . Tree farm committees, working with the 'Keep Green' movement [of fire prevention slogans] performed a signally successful task of indoctrination in a field marked by previous failures."[14] Aggressive firefighting programs developed for the Clemons were also a boon to area residents, whose towns and farms would not be in peril from the adjacent logged-off lands. Local residents had little to lose, and much to gain, from the company's fire control promises.[15]

The high cost of fire suppression forced Weyerhaeuser into an economic bind. To justify the extravagant cost of nurturing the plantings, Weyerhaeuser accountants stipulated that the eventual payoff in lumber sales had to be virtually certain. To make the fire protection system cost-effective, intensive artificial reforestation had to be undertaken in order to furnish something of value to be protected from fire. Indeed, establishing the infrastructure for the Clemons Tree Farm involved a large investment of funds by the company. Beyond the initial reforestation work, the company had to build and improve the site's roads, trails, buildings, and communication systems; buy firefighting and reforestation equipment and tools; and pay for fire suppression, debris and brush removal, continuing reforestation, and other maintenance once the Tree Farm was established. Three and a half months after the Clemons Tree Farm had begun, the project had already cost Weyerhaeuser over $56,000.[16]

Having made a financial investment in the Clemons Tree Farm, the Weyerhaeuser Timber Company needed to invest a proportionate amount to publicize it. It intended the Clemons to be not only an experiment in permanent forestry but also a public relations tool. Roderic Olzendam was in charge of organizing the dedication festivities. He told Phil Weyerhaeuser in a letter that he envisioned the event as an experiment in public relations. "I look upon this experiment as very promising from the point of view of public relations. I am hopeful that we can demonstrate that the people of the community will rally with enthusiasm and real interest around the constructive program of a private corporation. We have an opportunity here which seems to me to be unique." He emphasized that he would be in control of every aspect of the ceremony and of how the Weyerhaeuser name would be represented. He assured his employer that "all of the written material, including the advertisements, will be submitted to me before release."[17] The publicity surrounding the dedication of the Tree Farm included photos, advertisements in the regional newspapers, and, most impressively, an elaborate dedication ceremony to officially usher in the company's new undertaking. The total costs surrounding the dedication and its advertising were calculated at $1,114.18, a sizable amount for a ceremony that simply promoted the Weyerhaeuser name, without actually marketing any product or service of the Weyerhaeuser companies.[18]

The dedication of this first Tree Farm was a public relations spectacle. Its overarching strategy was to convince the public that Weyerhaeuser had their best interests at heart. In the short term, Olzendam said, the Clemons Tree Farm would be a well-managed, fire-free, taxpaying neighbor. In the long term, he said, the Clemons Tree Farm would supply work to the area's loggers and prove that the Douglas fir lumber industry was a permanent presence in the Pacific Northwest. The publicity aimed to show both the short-term and long-term benefits. The official naming of the Clemons Tree Farm took place in a ceremony at the Montesano Theater in downtown Montesano, Washington, on June 12, 1941. Hosted by Chapin Collins, then the editor of the *Montesano Vidette* and one of the Clemons Tree Farm's biggest supporters, the ceremony was a major event for the small town. Four rows of dignitaries sat on the stage, including Governor Arthur B. Langlie, the Grays Harbor county commissioners, the mayor of Montesano, and the head of the Montesano Chamber of Commerce. Also present were a number of Weyerhaeuser corporate officers along with the presidents of the local bank and insurance company. Many, including the governor, gave short speeches of praise for the Tree Farm and the Weyerhaeuser Timber Company's initiative in creating it. Following the speeches, the thirty-minute Weyerhaeuser promotional film *Trees and Homes* was screened. Before and after the speeches and film, a 35-person band performed, and the general mood of the proceedings was light and festive. The main goal of the evening was to promote the Tree Farm idea, not just to those in attendance but to the larger public as well. The proceedings were reported widely in the local and national press as well as in forest-related publications such as *American Forests*.[19]

Olzendam's comments at the dedication ceremony were aimed to stir the spirit of those in attendance, imploring them to respect the Tree Farm and understand the role of forests in all of their lives. He opened by saying that "'We the People' of Washington, every last one of us, separately, and all of us collectively—whether we fall or buck, saw or finish, teach or learn, enact laws or obey them, sell or buy, preach or practice, work or loaf, hunt or fish—each one of us has a personal forest objective. Each one of us wants to do something definite about *our* trees."[20] He suggested that the needs and expectations of all of the region's inhabitants could be met through the practice of tree farming. Olzendam's speech was not only praise for the Clemons Tree Farm, however; he addressed the critics of the lumber industry as well. While he acknowledged the importance of the existing national and state parks, he warned of the dangers of relegating any more forestland to recreation alone. If too much of the Douglas fir forest were preserved for recreation, he warned, Pacific Northwesterners would run the risk of becoming "a race of yodelers tripping around in feathered hats and slickers high among the 'rain forests'; there is even a possibility that fallers and buckers will be transformed into nursemaids to tourists from Tennessee; that our high climbers will become caddies, and our head sawyers may be serving tea and sandwiches to sightseers."[21] He urged the listeners to take pride in the lumber industry and to guard against allowing it to die out in favor of recreation and wilderness preservation. Continuing the message of his earlier speeches, he admitted to this audience that "we lumbermen of Grays Harbor County certainly do not claim that our stewardship has been perfect—far from it. . . . Richly endowed by the Almighty with natural resources, we of this County, and of most other counties, all of us have used those resources, shall we say, profusely? We have had a pretty swell set-up here, probably a better set-up than we could afford to keep up permanently."[22] The Clemons Tree Farm was the first step toward coming back to a realistic rate of consumption in the Douglas fir forest, a rate that could be maintained permanently.

Olzendam's presentation of timber cropping as a stylized, simplified process was easy to describe and understand, although in its simplification the economic and scientific complexity of such land management lost much detail and nuance. He reimagined post-clear-cut timber cultivation, remaking it from a rigorous and obscure aspect of forest management into a polished, easy-to-understand publicity strategy. With no professional background in either forestry or the lumber business, he pared away all the complexity from the mechanics of forestry

A tree planter in the Cedar River watershed, Washington State, demonstrates reforestation technique for the Douglas fir region. A one-person unit carried a supply of seedlings in a pail and wielded a grub-hoe for planting (1936). Photograph 95-GP-1956-367129; Records of the Forest Service, RG 95; National Archives at College Park, MD.

and simplified it greatly in the process of interpreting it to the public. His role at the Weyerhaeuser companies was to serve as de facto head of publicity for the Pacific Northwest operations, and for Tree Farms in particular. He managed the corporation's public relations in the region and supervised the development of Weyerhaeuser advertisements for both regional newspapers and nationwide publications like *American Forests* and the *Journal of Forestry*. When Olzendam spoke to logging communities in the Pacific Northwest, he told them they should value Tree Farms for their potential to furnish direct and indirect financial benefit to the communities. In a speech to the Pacific Logging Congress in Seattle in October 1941, he offered a framework for lumbermen to explain their industry to a doubting public. He stated that in the western United States and British Columbia, "there are 193 million acres of potential Tree Farms. . . . On those potential Tree Farms, in the shape of trees, are growing future materials, wages, security, taxes, dividends, freight, and money for research."[23]

As politicians debated increased state and federal regulation of the lumber industry, portraying a healthy, self-regulating lumber industry as valuable to local and national economic stability became crucial. Lumber executives realized that simply describing the benefits of Tree Farms would not be sufficient to garner support. Although the economic benefits of a reinvigorated lumber industry would be felt mostly in small rural communities of the Northwest, they also needed to win over those who lived outside the lumber-producing communities and rarely used the forests except for recreation. As western cities grew, along with urban interest in forests, parks, and wilderness, so did concern that the lumber industry was full of greed and waste. To win urbanites over, industry had to convince them that what they knew of the industry's profligate past would not be mirrored in its future activities. Thus, much of the campaign touting the bright new future of the lumber industry rested on presenting Tree Farms as a clean break from the industry's admittedly undisciplined past. Weyerhaeuser publicity materials often tackled the industry's poor reputation head on, as in, for example, the following passage.

> We do not claim that this program is perfect; we do not say that all of it has been put into effect by everybody in the industry by any means, but, as we attempt to add up our present situation, there is a clear indication that we are moving along the right line. . . . We of the industry welcome the constructive criticism of public opinion. A frank look at himself through the eyes of a friend is a good thing for any man. [We are] tolerant—open-minded and progressive. We do not claim perfection. We admit it would have been a splendid thing for our industry and for all concerned if the lessons we had learned during the past twenty years could have been learned forty years ago. But we are just average men, not any more far-sighted than other men in other industries.[24]

Olzendam both acknowledged the industry's past profligacy and emphasized its contrition to better emphasize the importance of Tree Farms to its current and future plans.[25]

Beginning in the days after the dedication ceremony, the Clemons Tree Farm was opened for any visitors who wished to drive out to the site. The on-site manager, Paul Meyers, was responsible for greeting and monitoring visitors as well as watching for fires and coordinating the fire crews, reforestation workers, foresters, and other employees working at the site. In the first six months of its existence the Clemons was a very busy place. The extensive regional and national press coverage of the dedication, as well as the Weyerhaeuser-funded publicity, made the site a regional tourist attraction for many forestry-minded Pacific Northwesterners. Meyers recorded 475 visitors to the site between the dedication in June and the end of 1941. Along with many local people from the Montesano and Aberdeen areas, a large number of regional luminaries made their way to the Clemons Tree Farm. These included Arthur B. Langlie, the governor of Washington; Compdon Wagner, president of the West Coast Lumbermen's Association; and Major D. W. Kent, district commander of the Civilian Conservation Corps. Many academic, governmental, and industrial foresters from the region as well as from across the nation also inspected the site, including Stephen Wyckoff, the director of the Forest Service's Pacific Northwest Forest and Range Experiment Station. Earle H. Clapp, then the acting chief of the Forest Service, visited the site during an official trip through the Northwest.[26]

The Clemons Tree Farm was not just a site for foresters and government officials to visit, however. Weyerhaeuser had acknowledged from the start the importance of incorporating the local people into the life of the Clemons Tree Farm. A notice it distributed to the people of Grays Harbor County read, "Fishermen, hunters and berry pickers are furnished with guest cards on request, except on days of extreme fire hazard, and are likewise asked not only to observe the state's fire laws, but to cooperate actively in promoting the success of a large-scale, unique experiment in practical forest growing. Visitors of every type are asked to remember that "Timber Is a Crop" and *Fire* is the worst enemy of that crop which produces payroll dollars."[27]

The invitation to the area's residents was, on one level, simply an extension of the public relations strategy that underlay every aspect of the Tree Farm project. Allowing recreational visitors would increase the ranks of the Clemons site's allies, since many local residents would benefit from it. However, the system was designed to closely monitor visitors. The manager inspected the guest cards and recorded all visits in a guest book. The issuance of a guest card was accompanied by admonishments about the risks of forest fires and reminders about the value, especially the economic value, of the Douglas fir seedlings on the Tree Farm. Such close monitoring of a forest in the Douglas fir region was extremely unusual for both public and private landowners. This strategy allowed the company to better control the land and the activities that took place on it and reinforced the assertion that the Clemons site was a farm as much as it was a forest. By allowing the local people to use the lands of the Clemons Tree Farm if they received permission, the company could better avoid the nearly inevitable trespassing and poaching. The guest card system in effect allowed a new sort of human control of events occurring on the Clemons, control that did not exist on most natural forestlands.

After the flurry of interest surrounding its opening died down, visitors continued to be drawn to the Clemons. Along with local residents who hunted, picnicked, or gathered berries, people from the wider region came to the

The owners of a non-industrial West Coast Tree Farm near Vernonia, Oregon, pose with their new West Coast Tree Farm sign. This rural family added tree farming to their ranch economy. The adoption of tree farming by rural inhabitants of the Douglas fir region influenced public opinion about the forest industry from the 1940s onwards. Courtesy Forest History Society.

Clemons as a drive-through destination on a weekend automobile excursion. The Weyerhaeuser Timber Company produced an informational brochure for auto tourists, available at the Clemons welcome center. The brochure guided tourists around a loop road through the Tree Farm, with descriptions of numbered sites of interest keyed to signs placed along the roadside. The focus of the brochure, not surprisingly, was to show the quality of Weyerhaeuser's forest practices and fire control. Most of the sites of interest highlighted by the booklet were specifically chosen as examples of the improvements the company's attentions were making on the logged-off land. The booklet contrasted areas that had been replanted by hand with those where natural reforestation had crept in. Even the sites where replanting had failed were portrayed as beneficiaries of the company's attentions. For example, Point 6 on the tour was described as "a foresters' 'problem area.'"

Repeated forest fires, which swept over this hillside before the Clemons Tree Farm was established, destroyed seed sources, and the land has not restocked satisfactorily. Some Douglas fir has established itself, and these scattered trees, now about 20 years old, are beginning to cast seed over the land. This may restock the hillside . . . provided that weed trees like vine maple do not choke out the seedlings. A Douglas fir will not grow in full shade. On research plots near this point, Weyerhaeuser foresters are seeking ways to bring such 'problem areas' back into full production.[28]

Because the road-building, replanting, and fire-protection work continued at a relatively fast pace, the views of the site were always changing. The most effective proof of the Clemons Tree Farm's value may have come several years after its establishment, when signage explained to visitors the speed at which the Douglas fir was regrowing in the replanted sites. As the landscape of the site transformed, the Clemons became a node of optimism about the Douglas fir logging industry as a whole.[29]

For all the on-the-ground effort to create the Clemons Tree Farm, its importance to Weyerhaeuser lay far more in its symbolic value than in its future profitability. The monetary investment in the Clemons tract was not very significant. While the company's investment in establishing the Clemons had run upward of $56,000 by the end of 1941, the net income of the Weyerhaeuser Timber Company in that same year had been over $8.5 million and its total assets were valued at well over $150 million.[30] However, the project was seen as important to the company not just for its short-term public relations benefits but also for its long-term profit potential. The company's annual report for 1941 described how these two goals worked in tandem. It stated that

the forest industries have stepped up their progress towards their own long-range objective of nation-wide timber cropping. . . . One over-all conclusion has been reached to date: Under favorable circumstances, which are attainable, privately owned forest lands can be managed to produce successive forest crops on a profitable basis, if coupled with public understanding and cooperation. At the present time public understanding lags behind the actual deeds of private land owners. Therefore, informing the public of our progress is an obvious responsibility. To do this it becomes essential to have definite measuring plots—areas upon which facts can be obtained, and where those facts can be demonstrated through visible accomplishments. The first such demonstration area of our company is [the Clemons Tree Farm].[31]

The company presented the Clemons Tree Farm as a small demonstration project, but one that had the potential to determine company practices and policies for the long term. Those involved in the formation of the Clemons Tree Farm had viewed it as a fairly large and complex undertaking. While small in terms of the Weyerhaeuser Timber Company's landholdings and finances, the Clemons Tree Farm project had an impact, both immediately and over the long term, that would go far beyond either its acreage or its cost.

Weyerhaeuser Timber Company promoted public exhibitions of its scientific prowess and land stewardship, even when it had little real substance to merit the reputation. Simultaneously with the opening of the Clemons Tree Farm, Weyerhaeuser invested $100,000 to add a research laboratory and full nursery at its manufacturing site in Longview, Washington. Executives pegged the Longview laboratory as "an excellent vehicle for doing a constructive public relations job" before the first scientist even set foot inside.[32] The Longview facility had already been touted as the Clemons site's source of fir seedlings and research scientists, while it was in fact barely operational as such. The company began constructing the laboratory in May 1942, skirting wartime construction restrictions because wood was a necessary war matériel. In June, with construction just under way, both J. P. "Phil" Weyerhaeuser Jr. and the future laboratory director, Clark Heritage, had already made public statements to the Longview newspaper about the lab's program of research. In late July, a reporter from *Fortune* magazine visited Frederick K. Weyerhaeuser at his Saint Paul office for an article about wood's potential as a "material of the future," and about corporate research at Longview in particular. Having no answer himself, he called Edwin Weyerhaeuser Davis, a company vice president, into the office. Davis recounted in an internal company memo that when the reporter "asked me what we proposed to do in the Longview laboratory and what fields of research looked especially promising to us," he was forced to admit that "the laboratory was not built, we had not hired any personnel and did not have a work program. I must admit my face was a little bit red."[33] By December 1942, construction was complete and the laboratory had eight employees on the payroll. However, the building still lacked sufficient electrical capacity to conduct planned research, so little scientific work took place.[34] Weyerhaeuser's West Coast public relations team nonetheless pressed headquarters to authorize a grand opening event similar to what had taken place at the Clemons in 1941. Those closely involved, cognizant of the lack of actual research on-site, worried about "beating the tomtom and inviting in outsiders" who would "ask what you propose to do in the laboratory, and answering that question might prove embarrassing."[35] Weyerhaeuser's stumbling blocks with the laboratory were almost entirely a result of the heightened war footing in the Northwest in 1942, and there is no evidence that it had set out to intentionally promote research that was not taking place. However, Weyerhaeuser executives clearly felt pressure to promote a positive image of their company as a cutting-edge, research-driven entity. Constructing an image as an enlightened, scientific company was far easier to accomplish than the hard work of conducting industrial forest research.

American Forest Products Industries and Tree Farm Promotion

The first Tree Farm did not change industrial forests on its own. Only when the idea of tree farming expanded to other corporations did it begin to transform industrial definitions of forests and forestry. Weyerhaeuser employees had generated both the forest management plan and the publicity machine to create the Clemons, but to make the Tree Farm take hold the Weyerhaeuser signature had to be minimized. Weyerhaeuser gave up much of its direct control of the Tree Farm idea, transferring it to what was at the time a little-known entity, American Forest Products Industries (AFPI). This group had been formed in 1932 to promote industrial research and commerce in the lumber and paper industries as a subsidiary of the well-established National Lumber Manufacturers Association. Founded in 1902, the NLMA was the foremost nationwide trade organization for the lumber industry and was active in lobbying and providing public information.[36] While its early projects had focused on research into new wood-based products, such as plywood and particleboard, by the end of the 1930s the AFPI was increasingly serving as the de facto publicity organ for the NLMA. At the spring 1941 meeting of the executive committee of the NLMA, Frederick K. Weyerhaeuser donated $300,000 to expand the AFPI. Fighting the negative public image of the lumber industry officially became the AFPI's responsibility, and the Tree Farm concept was a central element to that work. With the Weyerhaeuser seed money, the AFPI crafted a national campaign to redefine industrial forests both on the ground and in the public eye. Its main objectives were first "to make the American people aware of the fact that timber is a crop, which the forest products industries are endeavoring to grow and protect continuously," and second "to stimulate . . . further and continuing improvement in forest practices which are worthy of public approval." Its final objective was to explain that the goal of the program was not "to create in the mind of the recipient the impelling desire to immediately buy a given product" but instead to create "a long range confidence in the forest products industries."[37]

The Tree Farm program seemed to take off fast, but much of the acceleration was illusory. The AFPI certified dozens of preexisting reforestation projects in the months after the Clemons, exacerbating suspicions that Tree Farms would be of low quality and unreliable production. The AFPI did not require a site to have the same level of management as the Clemons to get certified, nor did Tree Farm certification mandate oversight by trained foresters. Many of the sites submitted for certification in the first years of the program had already been under some sort of industrial forestry management for a number of years. Certification simply redesignated existing reforestation as a "Tree Farm." The industrial forestry practiced on a site before its designation ranged widely in quality. Some were the sites of rigorous experimentation, such as the replanting trials on the Oregon Long-Bell lands mentioned in an earlier chapter. However, others had far less rigorous management, with little oversight or commitment on the part of the companies that owned them. Lumber companies submitted land for certification mainly for the public relations benefits they suspected it would garner them.[38]

The AFPI and the Tree Farm program challenged the authority of mainstream foresters. Many saw the programs as elaborate attempts to stave off public regulation of private forestry. The AFPI's certifications of Tree Farms were not based on sound forestry or done by respected professionals, and hence many foresters saw the designation as essentially meaningless. The Tree Farm certification was evidence not of a company's good forestry practice in general but only of the existence of a Tree Farm program on the company's land. A proposal for a Society of American Foresters "Seal of Approval for Acceptable Private Forest Practices," to be awarded by professional foresters,[39] underscored foresters' primary criticism of the Tree Farm. Beneath the AFPI's false professionalism, foresters criticized the program for being overly focused on the planting of trees while ignoring the rest of the spectrum of good forestry practice. Furthermore, it focused attention on the activities at a single site while ignoring the state of the rest of a company's lands and logging practices. By adhering to their own code of forestry conduct, lumbermen might be able to convince the public that government regulation of their practices was unnecessary. "No general statement can be made accurately to describe all units of the lumber industry," remarked an editorial on the Seal of Approval idea. Perhaps "the time may have arrived for foresters to think of the lumber industry not as something all good or all bad in so far as forest practices are concerned, but rather as a large number of individual units, some of which follow forest practices acceptable to the profession and others which do not."[40]

While professional foresters continued to object to tree farming's oversimplified approach to forest management, they could not prevent its rise in popularity outside the ranks of the profession. The AFPI first framed the term "Tree Farm" as a new class of forest use and then orchestrated praise and approval of the Tree Farms it certified. Having these dual responsibilities meant that the AFPI controlled the image of the Tree Farm completely. Although F. K. Weyerhaeuser's AFPI seed money had ostensibly been earmarked for broad publicity goals, the actual work of the 1940s AFPI was legitimizing the management plan of the first Tree Farms. Its reasoning behind certifying the Clemons Tree Farm as the nation's first was somewhat circular, since the certification procedure itself was designed around the management protocols of Weyerhaeuser's preexisting reforestation plan.[41] Weyerhaeuser executives wanted the Clemons to become a showpiece for industrial forestry, and its certification as "Tree Farm Number One" was intended to boost its public profile. The Tree Farm idea caught on quickly with industry leaders as they took note of the favorable publicity Weyerhaeuser had garnered with the Clemons Tree Farm. New, industry-friendly reforestation and forest tax regulations in Oregon and Washington also encouraged the formation of new Tree Farms in the region. The second Douglas fir Tree Farm to be certified was in Neah Bay, Washington, established by the Crown Zellerbach Corporation. The third to be certified was one established by the Long-Bell Lumber Company on its logged-off lands in western Oregon. All of the first eighteen Tree Farms to be certified by the AFPI were in the Douglas fir region and were planted with Douglas fir seedlings. The total overall area covered by certified Douglas fir Tree Farms by September 1947 was more than 2.5 million acres, held in fifty-one sites in western Washington and twenty-seven in western Oregon.[42]

Industry leaders hoped Tree Farms would illustrate their commitment to long-term forest health. Their attempts to do that through industrial sustained yield had been unsuccessful, but this simpler plan held more promise. Lumber companies did not intend their Tree Farms to be restored versions of the original forest but instead to be new landscapes designed to have the highest possible productivity. Because of this, company foresters often designed Tree Farms that strayed from the ecology of the natural forest. While inside the Douglas fir region most stuck to the Weyerhaeuser format of planting, in other forest types an improvement attitude often prevailed. Because of slow growth and small tree species, the sparse mixed-pine forests on the drier eastern slopes of the Cascades naturally yielded less timber per acre than most other western forest types. In 1943, Weyerhaeuser started the Klamath Falls Tree Farm near one of its largest and most modern lumber mills. At its inception, the Klamath site was the largest Tree Farm the company had established.[43] Although they planted ponderosa pine, one of the naturally occurring species on the site, company foresters planned this Tree Farm to artificially increase the eventual timber yield of the site.[44] By planting trees more thickly than normal and protecting the forest from natural fire and insect damage, both Weyerhaeuser and Forest Service foresters expected that the yield in the pines could be significantly increased. Weyerhaeuser foresters predicted that they could double their yield from logging the site's old growth when they eventually logged off the planted Tree Farm.[45]

The AFPI's Tree Farm program grew quickly, as the concept rippled out from the largest lumber companies to smaller operators and individual farmers and landowners. The AFPI evolved in several ways, accommodating growth through reorganization and the development of new programs. The AFPI officially separated from the NLMA in February 1946, allowing it to present itself as neutral and independent and to describe its focus as "solely with public information and forestry."[46] As a wing of the lumber industry's main trade organization, the AFPI and its programs were at risk of becoming an easy target for criticism from academic and federal foresters. Its income still came from industry sources, however, and its main task was still to publicize and promote the industry's best practices and most successful initiatives, including the Tree Farm. It still controlled Tree Farm certification and publicity and also promoted other industrial forestry projects. The AFPI developed and distributed educational materials that explained tree farming and supplemented the fire prevention and forestry programs in various states. As the scope of the Tree Farm program grew, regional committees of the AFPI took over more of the duties of certification and education. The AFPI empowered volunteer agents to inspect sites and review reforestation and protection plans. Landowners who demonstrated that they were working toward reforestation received AFPI certification, including a logo plaque to display at the entrance to the site.[47] The expansion of the Tree Farm initiative included the development of tree farming enthusiasm among smaller, nonindustrial landowners. Such Tree Farms, often established by crop farmers on parcels of unused acreage, were small but included the standard Tree Farm focus on commercial tree species and fire suppression. Certifying small Tree Farms not only ensured better planning and management on those sites but also allied their proprietors with industry and helped reduce the danger of fires originating on small landholdings. AFPI educational publications combined discussions of Tree Farm–centered forestry with a strong proindustry message.[48]

Beyond demonstrating a voluntary commitment to better forest practices, the term "Tree Farm" could also be expanded by the AFPI to include all sorts of woodlots and cutovers. Seeing the early success, the AFPI now broadened its definitions to make tree farming more appealing and attainable for lower costs than the original Weyerhaeuser Tree Farms. At the same time, however, the lumber industry and forest landowners did not suffer from any federal regulation of their activities. The NLMA policy stated that "this voluntary undertaking of private land owners, designed to assure continuous forest growth, has brought about a substantial improvement of forest practices and should be continuously expanded. . . . The cooperative and educational approach to forest problems should be more fully developed, but without Federal coercion."[49] The NLMA, the entity that had first held the responsibility of promoting the Tree Farm ideal in the early 1940s, had shifted most of the duties of overseeing the program to the AFPI when the two entities formally separated in 1945. However, the NLMA still worked to promote Tree Farms. In its Statement of Forest Policy issued in October 1949, the NLMA described the Tree Farms' importance to the industry's public image. It reads, in part, "Timber is a crop. When protected from fire, insects, and disease, and managed for continuous production, a forest area is a TREE FARM. We believe that the industry program of encouraging TREE FARMS is one means of assuring intensive protection from fire and the application of improved forest management practices on individual properties."[50]

These smaller companies and nonindustrial landowners relied on the AFPI to consult with and educate them on how to create workable tree farms. Too small to have in-house foresters and too cash-strapped to hire consulting foresters, these clients looked to the AFPI to take on the role of forester. While the largest companies cared more about improving their public image, smaller tree farmers often planted as an investment for future timber supplies. Without much oversight of calculations of planting regimes, much of their planning could be untrustworthy, even intentionally exaggerated to impress potential investors or others. These mixed motives worried the foresters at the more stable lumber companies just as they worried government and academic foresters. W. D. Hagenstein, a lifelong industrial forester and industry spokesman, explained the position in which he and other industrial foresters found themselves: "In deciding how we would use the Tree Farm idea to promote better forestry in the Douglas fir region we had many discussions on it as a public relations tool. Early in the program there was a tendency on the part of some Industry public relations men to claim better performance for Tree Farms than they were capable of attaining in those younger years. However, we soon calmed them down and convinced them that the only permanent public relations the Tree Farm program would yield must be based on solid performance."[51] While Hagenstein and others tried to counter the overoptimism of some tree farmers, expectations of Tree Farm yield within the industry remained quite elevated throughout the 1940s.

William B. Greeley, the former chief forester now overseeing Douglas fir industrial interests as head of the NLMA/WCLA, understood that controlling fire would require coordination between industrial interests, state and national forest managers, and the larger public. Effective cooperation began in 1940, when the state government of Washington, heavily influenced by pressure from Greeley, created a wide-ranging outreach program aimed to deliver messages about good forest stewardship to those who owned and used the state's forests. Keep Washington Green's campaign extolled the virtues of reforestation, forest fire control, watershed protection, and good land-use practices to all forest users, industrial and recreational. Much of the state program's initial approach was modeled on the public relations messages Weyerhaeuser and other companies had been fine tuning over the previous years. The campaign aimed to encourage the state's rural and urban citizens to respect and appreciate state, national, and industrial efforts to improve industrial logging practices. Keep Washington Green's strongest message was that fire prevention was every citizen's duty, which complemented and broadened the localized fire prevention efforts already under way in many public and industrial forestlands. Having indirectly set the tone for Keep Washington Green's messaging, lumber companies then presented Tree Farms as evidence that the industry was out in front on just the sort of forest stewardship that the state wanted. In a letter to Governor Arthur B. Langlie in April 1941, Phil Weyerhaeuser wrote that the Clemons Tree Farm was the "actual carrying out of what we conceive to be the responsibilities of a modern logging operator under the 'Keep Washington Green' program."[52] The company's motives for enforcing intensive fire control were to keep Douglas fir tree farming profitable, of course, but the Keep Green programs illustrate the interconnectedness of state and industrial goals. "We are inclined to believe," Weyerhaeuser wrote, "that this will constitute one of the most intensive reforestation and fire protection systems ever installed by a private operator in the United States,"[53] but wildfires obeyed no property lines. The Washington program was followed soon after by similar programs in Oregon and other forested western states. These Keep Green programs were also the immediate precursors to the Forest Service's widely heralded public campaigns against careless wildfire ignitions, including Smokey Bear. By imbuing a sense of forest stewardship and responsibility in citizens, both the public and industrial forest operators could save immense amounts of money on fire suppression. This state-financed, industry-inspired framework of fire prevention publicity helped reassure industrial landowners that investing in reforestation could be profitable.

Professional Foresters' Criticisms of Tree Farms

One of Weyerhaeuser's prime motivations for starting Tree Farms was to demonstrate that federal regulation of private forestry was unnecessary. In its 1941 annual report, the company acknowledged that the fear of regulation was a strong motivation, if not in the creation of the Tree Farm, then at least in the company's wide public promotion of it. It framed its argument in terms of public education rather than public relations, as was often the case with industrial foresters, stating that "the one effective weapon against a legislative blitz is a fully informed people. The job of giving facts about the forest industries to the people of the United States is the obligation of the forest industries and of the companies and individuals which make up those industries. Time is of the essence. There is still time, but it is later than we think."[54] Lumber companies had been under near-constant threat of increased regulation by federal foresters throughout the New Deal years. Many foresters read the industry's campaign cynically, as a way to continue voracious and destructive logging practices by hiding them behind statements about permanence and forest health. A *Journal of Forestry* editorial observed that while one of the Tree Farm program's purposes "is clearly to forestall federal regulation, the likelihood of which will be much increased unless private owners voluntarily adopt improved practices, [that] is certainly not to its discredit."[55] The existence of such an ulterior motive had little bearing on the efficacy of the program itself.

The Weyerhaeuser Timber Company had long had its enemies in the Oregon and Washington state governments, and its new Tree Farms did little to change that. Indeed, Olzendam's boisterous speeches about the Clemons had alienated many of Washington's foresters. The Washington State Forest Board developed an antagonistic relationship with the state's largest forestland owner. The board, which was charged with planning and land-use decisions in the state, consisted of T. S. Goodyear, state forest supervisor; Jack Taylor, state land commissioner; Edward Davis, director of the Department of Conservation and Development; and Hugo Winkenwerder, dean of the University of Washington School of Forestry.[56] In a frank discussion, the board members aired their suspicions and doubts about Weyerhaeuser motives, not just with Tree Farms but in all their dealings. The elaborate dedication ceremony for the Clemons Tree Farm seemed odd to many, and the constant pressure from entities like the NLMA reminded them that the company's true motive was simply profit.[57] The discussion took place only four days after the dedication ceremony, in response to a Weyerhaeuser petition to the board for approval of a land exchange plan with the Forest Service. The Forest Service wanted to transfer about eight square miles of federally owned forestland to Weyerhaeuser. In return for this land near the Wind River Experiment Station in the central Cascades, Weyerhaeuser would give the Forest Service some as-yet-undetermined lands elsewhere in the nation. Goodyear, the state forester, was dead set against such a proposition, since "the state and the county lose the land and gain nothing, while Weyerhaeuser gets some stumpage [permission to cut standing timber]." He worried that the company would not release its ownership of the lands after the cutting was complete. With Weyerhaeuser now claiming there was profit to be had in replanting and maintaining land for the long term, there was less reason for the state to trust that it would hold to its promises of relinquishing cutovers. Goodyear also spoke out against the entire mind-set of industrial forestry that had been aired at the dedication of the Clemons Tree Farm. He told the board, "I might say that the county commissioners turned it down. You recall their meeting [the dedication ceremony] at the Clemons Tree Farm the other night at Montesano and the remarks made about more national parks. [Returning this land to the state] is hardly consistent with the company's policy. It just gives them a chance to pick up a little more stumpage."[58] Olzendam's disparaging remarks about the importance of public forestlands alienated foresters who dedicated their careers to those lands. At the Clemons dedication, Olzendam had implied that the creation of any new parks, wilderness areas, or other forest designations where logging was prohibited would hasten the downfall of the region's lumber industry. He had depicted state land management priorities as an economic threat, saying that "of course, we must have plenty of places to play, but, if we insist upon fencing off more and more *relaxing areas* in the form of National and State Parks, there is danger" that the industrial base of the region would begin to decay.[59] Goodyear swayed the board, which voted unanimously to deny the company its request.

The Washington State Forest Board's decision illustrates foresters' mistrust toward Weyerhaeuser and other large lumber companies during this era, and their attempts to push back against the increasingly powerful industry. The board suspected that Weyerhaeuser, even more than other lumber companies, was generating vast profit by unfairly exploiting the state's resources. Even when the company logged on its own lands, the state ended up underwriting costs, especially through road building and maintenance. If the main purpose of the roads was the removal of logs by private companies, the state was loath to finance them, even on state-owned lands. But if a road had value as fire protection, the state was more likely to fund it. As fire protection strategies developed, saving privately owned standing timber from fire became an increasingly expensive state responsibility. For example, the Weyerhaeuser Timber Company asked the state to build a single stretch of fire road across a portion of state lands that bordered the company's forests but also asked it to waive the customary road-building and right-of-way fee because the road would help with fire protection. Within a week of the acquiescence to that request, Goodyear saw "a flood" of applications from the company for building more free roads to access various parts of its land.[60] Goodyear told the board that "I do not think that some of these applications are made entirely for fire protection purposes, but rather as a short-cut into some of their [logging] camps."[61] Weyerhaeuser hoped to log right up to the edge of state forest and then transport timber the short distance across state land on these fire roads, rather than having to bear the high expense of building logging roads through its own land. Conversely, Goodyear speculated, if a situation arose where the state needed to build a road over Weyerhaeuser's lands, "if we did have reason for having a right of way over their land you can be assured they would charge us for it."[62] The board began denying all requests by Weyerhaeuser Timber Company for free roads unless they were needed for fire protection. However, they did move to build such roads in cases where they would be of great use for fire protection, even if those roads would also be of great use to the company for transporting logs and logging equipment. The interests of the state forester were clearly at odds with those of the Weyerhaeuser Timber Company, yet the company still benefited from the grudging help of the state.

Public approval did not mean Tree Farms were good forestry, only that they were effective publicity. While industry executives were immediately excited by the simple yet evocative image of the Tree Farm because they welcomed any innovation that might improve their image, the foresters they employed were more reluctant to accept it and worried that the plan seemed superficial and unsophisticated. Industry forester W. D. Hagenstein immediately disliked the Tree Farm idea. He recalled that "as a forester I didn't react favorably at first to the term, 'Tree Farm,' because it sounded like we were classifying ourselves with the cotton cultivators, radish raisers and punkin [sic] pickers and somehow, no self-respecting logger wanted that." He changed his mind after seeing that the public embraced the Tree Farm idea: "After much exposure to the idea and talking to people in all walks of life all over the Northwest, I soon found out that people did not understand what forestry was all about from our glib use of 'sustained yield,' 'forest management' etc. Therefore, I realized Tree Farming was a useful tool and buckled-down to promoting it enthusiastically. I used the term in every way I could to promote better forestry and better understanding of what forestry was about both within the Industry and out."[63] While this could be seen as a triumph of public relations over forestry, the lumber industry seemed so strongly to need a positive public relations strategy that foresters like Hagenstein put their misgivings aside. Agricultural imagery described in accessible language made the project more comprehensible to the public than had the complex reckonings of sustained-yield management. Industrial foresters found themselves compromising their professional integrity to adopt the format favored by their employers.

As the Tree Farm focus pulled industrial foresters farther away from previous priorities, they trained their attention even more on the money tree instead of the whole forest. Note that the term used was not "forest farm," but "Tree Farm." The purpose of the Tree Farm was not to re-create the forest that had produced the trees but simply to reproduce the trees. The first generation of foresters in the Douglas fir had thought that the best way to regenerate the forest was first to understand its basic biology and then to re-create the processes that gave rise to natural forests. Instead, the Tree Farm was based on the preconception that the only desirable facet of the Douglas fir forest was the Douglas fir itself. The Tree Farm was an agricultural endeavor in that its goal was the production of a large harvest of a single product, Douglas fir lumber. By adopting the agricultural template for forest production, the industry privileged the processes that led to the establishment of its intended species. By planting Tree Farms, Weyerhaeuser and other lumber companies were committing themselves to a long-term future. Unlike most other industries, the lumber industry had to plan generations ahead in order to secure its long-term health. The goal of the Tree Farm was simply the creation of a future supply of lumber. While the Tree Farms might be used for other purposes, such as hunting and sightseeing, these activities would always take place within the larger framework of timber production.

Federal Forestry and the Role of Industry

Among professional foresters, the reaction to tree farming was mostly negative. Many saw the Clemons Tree Farm as solely a publicity stunt on the part of the Weyerhaeuser companies. Without knowing the extent of the reforestation or the techniques by which it was being done, many foresters rejected the industry-funded and industry-approved project of the Tree Farm on the basis of its context alone. Many academic and governmental foresters interpreted the christening of the site as a "Tree Farm" as merely an attempt to stave off the threat of public intervention in the industry's activities. Underlying these suspicions was an academic preconception that industry-employed foresters never had any real opportunity to practice the science of forestry and were kept on by lumber companies mainly as a palliative to criticism. Indeed, since the Progressive Era, there had been attempts to force implementation of good forestry practices on the lumber industry. And for almost as long, industry leaders had been insisting that such intervention was unnecessary. While there were innumerable instances where lumber companies simply logged the trees and then abandoned a devastated site, industry leaders claimed that they were already practicing good forestry when it was economically possible. The Clemons Tree Farm had been the industry's most successful attempt to date to convince the public that they were good stewards of the forest. Perhaps because of that success, mainstream foresters assumed the Tree Farm was a fiction, persuasive but not trustworthy.[64]

An editorial in the *Journal of Forestry* summarized academic foresters' ambivalence toward the Tree Farm concept. It acknowledged the frustrations many foresters felt with the style of presentation that Olzendam and the other public relations experts used in advertising the Tree Farm to the media and the public: "Some may be irritated at the occasional suggestion by overenthusiastic spokesmen for the movement that they are the first discoverers of the fact that trees grow. Others will regret that the sponsors of the movement felt it necessary to invent a new term to describe practices which in fact and by definition are actually forestry. Still others will resent the implication that 'common sense' is a novel characteristic of Tree Farming as contrasted with other forestry."[65] In the end, however, this editorial concluded that the Society of American Foresters should welcome any improvement in private forestry practices. It pointed out that for years academic forestry had not been greatly concerned with the economic aspects and profit possibilities of reforestation. Reforestation efforts had been conceived of in the pages of the *Journal of Forestry* largely as a remedy for wrongs done to the land through lumbering or fire. Most academic foresters had imagined the product of reforestation to be a replacement of the original forest. The new industrial foresters, however, were projecting their reforestation efforts one step further into the future. They conceived the product of reforestation to be not a replaced forest but instead a second crop of lumber from the site. The editorial acknowledged that many professional foresters had neglected the fact that the lumber industry was driven by economic concerns rather than either an aesthetic or ecological view of the forest. It stated, "We must admit that as a profession we have not been conspicuously successful in presenting forest production as a business enterprise in such a way as to command public understanding or to induce its general practice by private owners." If the AFPI could do this through the Tree Farm program, then perhaps foresters should give the program their blessing.[66]

Mainstream foresters were not only put off by industry's focus on publicity but also dubious that the planting regimes would lead to much more success. Several years earlier, as the agricultural language of industrial forestry was catching on, foresters had objected, and now those objections were revisited. The main concern for academics was that Tree Farms were "disregarding the many other products and services with which forestry is concerned, such as forage, wildlife, recreation, amelioration of climate, protection of the water supply, and prevention of erosion."[67] To academic foresters, it appeared that the Tree Farm concept was taking silviculture, one element of the larger academic forestry curriculum, and allowing it to stand in for forestry as a whole. Artificial reforestation and fire suppression would not of themselves create a forest. Ignoring soils, hydrology, and ecology seemed to portend an unsuccessful future for the plantings. The abbreviated descriptions offered by journalists compounded professional foresters' mistrust, since they simplified the project further for their readers. The experimental work planned on large tree farms like the Clemons was often misrepresented, to the point that the essentials of forestry practice were also misrepresented.[68]

Lyle F. Watts was appointed chief of the Forest Service in 1943, and one of his main activities in office was to push back against the lumber industry's growing power to set agendas in timber country. Watts campaigned against the Tree Farm movement on the grounds that it was a superficially appealing program created by untrustworthy companies and thus might undermine support for federal forest policy. Watts was part of a chorus of professional foresters who were highly doubtful that the technical aspects of Tree Farm forestry were sufficient to create good and lasting forests.[69] He warned that although the acreage of designated Tree Farms was growing rapidly, "unfortunately, mediocre or lower performance has served to qualify some properties for the 'Tree Farm' designation."[70] Against the backdrop of wartime demand, Watts even resurrected the specter of timber famine, warning that forests were not supplying enough timber products and hence regulation and conservation were still necessary.[71] Watts critiqued the Tree Farm program and the lumber industry to audiences of both professional foresters and the general public. Other federal foresters joined him, with assistant forester C. Edward Behre writing that the Tree Farm program was "so questionable as to constitute a challenge to the forestry profession."[72] For Forest Service foresters, public focus on the Tree Farm idea obscured larger problems and distracted people from the mismanagement of lumber resources. They argued that Tree Farms and other industrial forestry programs did not do enough to prevent future shortages or price fluctuations in lumber. The industry's assurances were based on faulty initial assumptions, they said, and the extensive publicity surrounding the Tree Farm idea only proved that the endeavor was merely industry propaganda.

The Sustained Yield Forest Management Act became law in 1944 and ushered in a new era of forestry. Like the O&C Sustained Yield Act before it, the 1944 act was heavily supported by industrial forestry groups. Also like the O&C Act, its provisions were based on David T. Mason's formulation of sustained-yield forestry. The goal of both of these acts was to economically stabilize a particular lumber market or region by controlling the rate of cut. By permitting only enough logging in an area to offset the rate of the forest's regrowth, the managers would ideally create a state of continual productivity instead of boom-and-bust cycles, benefiting local communities and large, well-established lumber companies. Some foresters criticized the terms of the act as catering to the desires of industry. Lost in the sustained-yield equation, other critics pointed out, was the ecological well-being of the forest itself. For some federal foresters, the most disheartening part of the act was that Congress had usurped the Forest Service's role. Established to both control land and manage the forest growing on that land, the Forest Service had now been forced to cede some of its authority in public forest management. On the other hand, some federal and academic foresters heralded the 1944 Sustained Yield Act for its technical sophistication and saw it as a rebuff against the Tree Farm movement. Federal sustained yield had the potential to rationalize the nation's lumber industry and create stable communities across timber country.

The 1944 act centralized management decisions and standardized logging policies on public lands. In effect, it heralded the beginning of a new industrial era for American forest management. Because this legislation created a production-centered agenda for most federal forestland, it also pushed the profession of forestry's focus further toward industrial production. Federal sustained yield and tree farming were two major developments for forest management in the early 1940s, and both focused on harvesting a steady supply of merchantable timber, not on maintaining overall forest health. Foresters had raised alarms about the simplistic Tree Farms, warning that they would not furnish enough timber to satisfy future demand. The Tree Farm program might compete for attention with sustained-yield forest management, but its acreage was still small. Federal sustained yield, on the other hand, encompassed large areas and multiple stakeholders. The sweeping breadth and long range of federal sustained yield, seen first in the Sustained Yield Forest Management Act, would transform American forestry as well as American forests.[73]

Conclusion

The Weyerhaeuser Timber Company and its affiliates significantly affected the course of American forest management in the early 1940s. First with its "Timber Is a Crop" campaign, and continuing with the development of the Tree Farm, Weyerhaeuser worked to drive public opinion toward a more tolerant view of industrial lumber operations. New Deal politics had created an atmosphere antagonistic to industry, with increasing regulation and oversight. With the success of massive projects like the Shelterbelt Project and the Tennessee Valley Authority, by 1940 even the possibility of nationalization of private forestlands had not been too far-fetched a fear within the industry. James P. Selvage, a prominent public relations executive, later reminisced to Frederick K. Weyerhaeuser that in 1940 and 1941, "the newspapers and magazines were flooded with material antagonistic to the forest industries," topped with the "completely alarming" depth of animosity and misunderstanding uncovered in the 1941 Opinion Research poll, discussed in chapter 5. Selvage recalled how, in the wake of that poll, Weyerhaeuser's efforts to change public perception had resulted in "thousands of columns in almost every publication in the United States seeking to allay the fears of the public that there would be a shortage of wood," efforts Selvage described as "an attempt to 'take the hysteria out of the public mind.'" He reflected that "this was the most successful public relations campaign about which I have ever known. The railroads have spent millions of dollars in both publicity and advertising and still are under threat of government ownership. But, with the expenditure of only a few hundred thousand dollars, the forest industries at the end of four years found the 'hysteria' gone."[74]

Indeed, public perception of the lumber industry changed significantly between 1940 and 1944. This chapter has focused on the development of the Tree Farm as both an environmental management scheme and as a landscape-scale public relations effort. The Tree Farm project took place against the backdrop of World War II, when public sentiment about many New Deal programs changed. During years of such great geopolitical and economic change it is hard to follow Selvage's lead and heap all the credit at Weyerhaeuser's feet. The Tree Farm did, however, offer a seemingly viable alternative to government regulation or sustained-yield forest management. It was more than a simple public relations ploy. Indeed, professional foresters in the Forest Service and elsewhere took Tree Farms seriously, subjecting the program to intensive scrutiny and criticism. The enactment of the Sustained Yield Forest Management Act, however, had an even greater impact on long-term forestry agendas. Federal adoption of sustained yield began an era of forest management in which economic considerations were of utmost importance in all decision making. Professional foresters both in and out of the Forest Service would be affected by the changes to come, as the political and economic agenda for western forests shifted. While the fight over Douglas fir would continue, the terms of the battle would change significantly.

Conclusion
Axe in Hand

I have read many definitions of what is a conservationist, and written not a few myself, but I suspect that the best one is written not with a pen, but with an axe. It is a matter of what a man thinks about while chopping, or while deciding what to chop. A conservationist is one who is humbly aware that with each stroke he is writing his signature on the face of his land.

Aldo Leopold, "November: Axe-in-Hand," from *A Sand County Almanac*[1]

In "Axe-in-Hand," Aldo Leopold meditates on the necessity and the process of cutting down a birch tree in his forest and muses over the reasons why he has chosen to cut a red birch rather than the white pine next to it. The reasons, he thinks, behind his decision to cut a tree are far more difficult to explain than they might seem on the surface. He might be motivated to manage the health of that stand of trees, deciding to cut one tree to create open space in the canopy, or to prevent a preferred species from being crowded. His choice might be motivated by a desire to possess a wilder forest, if one tree is better than the other at nurturing rare wildflowers, feeding wildlife, or fostering biological diversity. His reasons might be sentimental: "I planted the pine with my shovel, whereas the birch crawled in under the fence and planted itself." Or they might be practical: since the "pine will ultimately bring ten dollars a thousand, the birch two dollars," he asks, "have I an eye on the bank?"[2] Leopold could, if a neighbor had strolled over to inquire as he chopped down the tree, have provided any one of a dozen reasons why he was cutting the red birch tree, all of which could have been believed. But there is a deeper, less easily explained reason he cut the birch and spared the pine. As he put it, "Is the difference in the trees, or in me? The only conclusion I have ever reached is that I love all trees, but I am in love with pines."[3]

For a person standing in the woods, deciding which tree to cut, forestry is a practice, not merely a profession. "Axe-in-Hand" can be read as an illustration of the limits of theory and training in guiding one toward nurturing the health of the land. Leopold, himself a trained forester, illustrates the paradox of managing forests—any person who walks into the woods carrying axe in hand also carries

Aldo Leopold examining a young red pine at his "Shack" near Baraboo, WI (1946).
Although he resigned from the US Forest Service early in his career, Leopold maintained
a keen interest in national forest policy, the forestry profession, and, indeed, in the pines
growing on his own sand county land. Courtesy Aldo Leopold Foundation.

an entire host of motives. Leopold knew that in practice, forest management
was hardly objective but instead crowded with individual agendas, unspoken
biases, and contradictory ideals. He presented idealized, holistic visions of land
management elsewhere in *A Sand County Almanac*, especially in the "The Land
Ethic," the essay that included the "A-B Cleavage" discussed in the introduction
to this book. In "Axe-in-Hand," though, he presents a counterpoint to that ideal:
the messy reality of on-the-ground forest management, full of self-doubt and
subjective choices. If a lone tree cutter could have so uncertain a vision, it should
be no surprise that when many people hold that proverbial axe simultaneously,
agendas blur and fights ensue. This book has examined how those charged with
making decisions about Douglas fir have understood and evaluated both the
forests and the motives underlying forest management.

In the early-twentieth-century American Douglas fir forest, there was no clear
demarcation between wild and tame. The region's first generation of foresters

soon found that imposing some sort of order, both on the forest itself and on those who were harvesting it, was a daunting task. By 1910, a decade after Frederick Weyerhaeuser's decision to move much of his business west, the Douglas fir forest had begun transforming into a lucrative zone of industrial production. No observer of the region, whether forester, lumberman, citizen, or politician, could deny the breadth and intensity of change brought by market capitalism paired with large-scale, technologically advanced logging. The rapid growth of logging in the region had created significant ecological problems by the 1920s, coupled with the economic stress of an unstable industry. The explosive growth of the region's logging industry and its effects on policy and economics prompted America's foresters to scrutinize their profession's aims. American forestry, they realized, could not fruitfully proceed along the path sketched out by their European intellectual forebears. American foresters' attention focused on the specific problems of designing forestlands robust enough to maintain a permanent lumber industry, but it broadened to encompass ecological, recreational, and aesthetic aspects of forest management as well. The specific challenges and paradoxes of working in the Douglas fir region inspired a more widespread introspection about America's forestlands. Foresters raised concerns not just about the direction of decision making in the Northwest but about the values and assumptions underlying forest science and policy. Foresters, lumbermen, activists, and policy makers all looked for permanent solutions to the region's growing environmental instability.

Foresters' struggle to define the real value of their work led to debate over the utility and value of forests. Before the late nineteenth century, almost all American forests had, of course, been multiple use. In the Douglas fir region, humans had dedicated small areas to settlement, food production, hunting, and leisure, and these intermingled and overlapped with one another and with logging zones.[4] As forestland ownership in the Northwest consolidated under governmental oversight, this shifting patchwork became more stable, with large swaths of forestland managed for single purposes. While much of the Douglas fir forest was on public land, under a variety of federal and local designations, industrial ownership was also important. As the lumber industry grew in size and influence in the Northwest, ever more forestland was designated as open for lumber production. By the 1940s, most forest managers, in both government and industrial employ, approached and understood Douglas fir forests primarily as sites of continual lumber production. Many of the era's wilderness activists adhered to this approach as well, promoting protected zones of wild forests within a region mostly given over to industrial activity. The managed industrial forest was seen as the norm, with ecological complexity of secondary importance to timber productivity. Many of those who found this new hierarchy objectionable discovered a better home within the growing ranks of ecologists. By the end of the 1940s, death and attrition meant that fewer dissenting voices were heard within professional forestry circles. The adoption of federal policies of sustained-yield management solidified the profession's focus on perfecting methodologies tailored to accommodating permanent industrial presence.

Sustained Yield and After: Forest Management since 1944

There are different ways to value the trees in a forest, as Leopold reflected in "Axe-in-Hand." The adoption of permanent management programs in the form of sustained-yield management and tree farming marked the beginning of a new stage of professional forestry. When forests were seen primarily as a generator of logging revenue, the space allowed in the profession for ecological research, recreation, and preservation diminished. As federal agencies became more dependent on the revenues from logging, they worked harder to make logging lucrative for the companies involved. The industrial view of Northwest forests underpinned both federal forests and large-scale private forests for many decades. On the other hand, forest ecology, alienated from forestry in the 1930s, rose to prominence within the growing discipline of ecology. The scientific premises of ecology would also inspire innovative management approaches, led by the "Land Ethic" that Leopold articulated in the posthumous *A Sand County Almanac*. The Land Ethic, and other articulations of ecological management in midcentury, proved early mileposts in the development of myriad forms of sustainable resource management. Currents of change in forestry reflected not just what was at stake within that profession but also how stakeholders nationwide tackled the challenges of managing lands and resources.

Evidence of the industrial transformation of Douglas fir forestlands was unmistakable and widespread by the end of the 1940s. During World War II, the Forest Service had been compelled to furnish wood for defense purposes, mainly as building materials. The agency responded to the increasing demand for wood by authorizing more logging in the national forests, especially in the Northwest. Postwar, as military personnel returned home and travel restrictions lifted, recreational access became an important consideration in most national forests. In the Northwest, however, recreation remained largely secondary to lumber on national and state forestlands. The Forest Service found the new regulatory framework provided an opportunity to raise significant income through timber sales. Foresters who already championed industrially focused forests found that the revenue not only benefited the agency but swayed politicians toward intensive exploitation of publicly owned Douglas fir.[5]

For industry, the changes were visible on the horizon, although not yet an ongoing concern. In a 1951 report, Weyerhaeuser Timber Company stated that "in 1950, after half a century in the timber business, virgin old growth timber is still the primary source of logs for Weyerhaeuser's eight manufacturing centers. These forests will continue to supply high grade logs to company mills for the next several decades."[6] In this logging-friendly atmosphere, lumber companies stepped up clear-cutting on their own land as well as on public land. Advances in heavy-equipment technology meant that once equipment was in place, it was feasible to cut far more timber with far less labor than before. The largest companies also continued their practice of establishing Tree Farms on cutover lands. In the above-quoted passage in the 1951 Weyerhaeuser report, the observation about the continued reliance on virgin old growth is immediately followed by the comment that "gradually and inevitably the young timber crop now growing on tree farms will become the primary source of logs for pulping and sawmilling."[7] By midcentury, individual companies and lumber industry trade groups had crafted an image of the forest as the domain of loggers and a place of industrial production. Fearing taxes and regulation, they crafted a message that the forest industry was a boon to the region's overall economic health. This concerted effort to mold public opinion benefited industry twofold, in that it created a public both tolerant toward industrial logging on public lands and trusting toward the industry's activities on its own lands.[8]

The fate of wild forests changed too. As discussed in chapter 5, as Marshall mapped out The Wilderness Society his vision of forest wilderness depicted an ecologically whole place where no human activity—not even logging—was banned outright, but where modern convenience and technology in any pursuit was anathema.[9] After his death, Marshall's complex and inclusive definition of wilderness was replaced with a more concrete one, easily grasped and easily defended. As The Wilderness Society's leadership refocused its message on the recreational value of wilderness in the 1940s, the organization gained strength and visibility. Its advocacy for the development of federal wilderness policy succeeded with the passage of the Wilderness Act in 1964. The act established the basis for the designation of wilderness areas, many of the first of which were in the western forests of the Rocky Mountains and the Cascades. The aim was to preclude development of choice parcels of federal lands not yet heavily modified by human use, often because of their remote, rugged, or high-altitude locations. These federal wilderness areas completely excluded commercial or industrial activities, even on the smallest scales.[10] The new Wilderness Act defined wilderness only as a place where a person "is a visitor who does not remain,"[11] precluding anyone from making a home or livelihood therein. This definition highlighted the aesthetic and recreational use of the wilderness, which humans visited merely for pleasure. Although the Wilderness Act seemed to some to favor sublime peaks and canyons, the primeval wilderness of the forests still mattered. As Marshall had argued, the beauty of ancient forest was exquisite in its own right, and worthy of saving. The simple definition of wilderness at the heart of the new act meant that selected locations would be subject to strict guidelines and distinct boundaries. The federal wilderness system's emphasis on the absence of human presence cemented and validated ecologists' preexisting inclination toward the study of ecosystems free of the variables introduced by human presence.[12]

The vulnerability of untouched forests within the Douglas fir region had been a primary motivation for the Zon coterie's interest in wilderness. These forests remained under intense pressure to be logged as the acreage of old-growth Douglas fir dwindled. In Washington, lumber companies had appropriated the heavily forested edges of the protected Olympics piece by piece. In the early 1950s, the recreational and aesthetic value of Cascades forests was assumed by the Forest Service to be secondary to that of the high peaks. In Oregon, the Forest Service proposed redrawing the boundaries of the Three Sisters Primitive Area so that a low-elevation swath would be opened for logging. As the power of the region's lumber industry grew, however, pitched battles were fought over the limits of industrial growth in the region. Activists held that it was vital to save not just the rugged mountain peaks above treeline but the thick forests clinging to the flanks of those mountains as well. Indeed, the high peaks were so difficult to access that even if they had held valuable resources, the costs involved were prohibitive enough to deter development. Ancient Douglas fir forests, on the other hand, were no less precious and were extremely vulnerable to destruction. National wilderness legislation, then, needed to be strong enough not just to protect peaks but also to armor ancient forests against onslaughts from the lumber industry. Thus, protecting these last intact forests from industrial activity became a priority. Within the Douglas fir region, several wilderness areas in the high country of the Cascades were established within the existing national forests after the passage of the act. The wilderness areas joined a very small number of other large Douglas fir forests, mainly in national parks such as Crater Lake, Olympic, and Rainier, where ordinary logging activity is now barred. Besides these large protected swaths, small remnants of ancient forest may be found throughout the region on public and private lands. The era of unmanaged, untamed, ancient "old-growth" forest in the American Pacific Northwest, however, has long since departed.[13]

The science of ecology grew in prestige and sophistication in the post–World War II years. From its roots as an innovative new approach within established biological disciplines, it grew into an academic discipline unto itself. Ecology's intellectual ties to forestry became weaker, as did its ties to agriculture and other fields of land management. Those who identified as ecologists became less interested in promoting the utility of their work as the field became more intellectually secure and institutionally independent. As the field reached professional maturity, the growing ranks of ecologists established more ambitious and complex research programs than ever before. The scope of the discipline grew with the help of the midcentury increases in funding for basic research from federal agencies such as the National Science Foundation. Studies of the dynamics of living things in their natural settings were soon structured primarily with ecological techniques, vocabulary, and approaches. As ecology's star rose in the 1950s and 1960s, forestry remained focused largely on management of industrial forests. The relative lack of ecological research or discussion within professional forestry became more obvious as ecology permeated other biological fields. This book has shown how the profession of forestry alienated the ecologically minded from its ranks. By the late 1950s, the two disciplines had little interchange or communication, and ecology's rising prominence within the biological sciences did not immediately affect the conduct of forestry.[14]

Ecology changed not only the intellectual terrain of science but also public discourse about environmental issues. Esteem for science and technology reached new heights during the Cold War years, in American mainstream culture as a whole and in environmentalism in particular. A new strain of environmental thought, invigorated by scientific insights into the interconnectedness of all living things on Earth, emerged in the early 1960s. While earlier environmentalist discourse had focused largely on the preservation of wild lands and species, this new second wave also incorporated urban sanitation, industrial waste, and other previously neglected issues. Rachel Carson's best-selling *Silent Spring*, published in 1962, used the language and sensibility of science to decry devastation of natural systems through chemical contamination.[15] Activists and concerned citizens advocated for the health and stability of entire ecosystems and pressed federal and state agencies to incorporate more holistic interpretations of nature into wilderness designation and natural resource decision making. The inflexible top-down planning procedures of the Forest Service, rooted in sustained-yield and multiple-use management provisions, could not accommodate such pressure easily. Environmentalists targeted the Forest Service for criticism, as they saw its foresters as catering to industry desires for low-cost logging contracts on public lands. Underscoring activists' assertion was the demonstrable economic benefit both the agency and the industry reaped from this arrangement.[16]

The larger conflict between ecologically minded environmentalists and the Forest Service was mirrored in conflicts within the professional confines of academic forestry. Much of the research funding and employment opportunities for American universities' forestry departments depended on the lumber industry, either directly or indirectly. Thus, the academic discipline of forestry remained tightly linked to industry, just as the Forest Service did. However, new ecological methods did find their way into some foresters' research by the 1960s, reflecting the explanatory power and intellectual appeal of ecology apparent throughout academic biology. The career of Jerry Franklin provides an exemplary case of the emergence of ecological approaches within academic forestry. Franklin, training at Oregon State University's School of Forestry in the late 1950s, had determined early in his career that the ideas and methods of ecology could offer him great new insight into questions about Northwest forests.[17] Indeed, Franklin founded his career on basic research in forest ecology, becoming the first scientist to undertake a major ecological study of intact Douglas fir forest. By 1969, Franklin had secured National Science Foundation funding to include Oregon State University's experimental forest as a research site as part of the United Nations' International Biological Project. Franklin took on increasingly complex and sophisticated ecological studies of the Douglas fir region over the course of his career, often funded by the National Science Foundation.[18] He rejected the premise that a living forest was simply a supply of marketable wood not yet freed for harvest. Reflecting on the trajectory of his career, Franklin stated, "I got a heavy-duty education at Oregon State in the traditional view, that there was value only in wood and that these forests needed to be converted. The idea was that these were really negative ecosystems . . . [but] I thought there's got to be value in there. They can't be negative ecosystems—it's got to be the way we're looking at them."[19] Except for Franklin and a handful of other mavericks, basic ecology had little representation within schools of forestry during the 1960s and 1970s. In addition, that era's academic ecologists focused mainly on the study of pristine and remote ecosystems, where human impact was limited, along with controlled experimental manipulations.[20] Studying a highly human-impacted forest was also impractical, as quantifying the extreme complexity in such a landscape was difficult with the tools of analysis and measurement ecologists then had available. Even after the passage of ecology-centered environmental legislation in the 1970s seemed to herald ecology as the new worldview, academic forestry held academic ecology at arm's length. True ecological forestry remained an underdeveloped concept, isolated from the profession's real centers of power and money.[21]

In the 1970s, new frameworks of environmental management emerged via the passage of environmental legislation and the establishment of new oversight agencies. The incorporation of the main ideas of scientific ecology into an increasingly energized environmentalism precipitated a wave of reform that leapfrogged the old-fashioned Forest Service entirely. The overall aim of this era's environmental laws and regulations was to safeguard American nature by protecting not just individual species or particular landscapes but entire webs of ecological interactions. The tide of change in this era included the creation of the Environmental Protection Agency, the passage of the Clean Air and Clean Water Acts, the creation of the Superfund program, and other legislation and regulation. The effect was to create, in a relatively short period, a new body of interlocking and ecologically informed environmental restrictions. The Forest Service's industry-friendly policies conflicted with the more committed and integrated ecological mind-set embodied in these new governmental entities. The agency was forced to change its standard practices in order to come into compliance with the government's new rules and initiatives. Environmentalists, and others who had been rebuffed when facing the Forest Service directly, could now address demands straight to the newly created regulatory bodies. The Forest Service, which had resisted incorporating a more ecological worldview into its practices, now often found itself compelled to do so.[22]

The Bitter Ends: The O&C Revested Lands in the Twenty-First Century

The fight among forest professionals over Douglas fir was a struggle to determine how and where value was found in the forest. By locating and defining this value, the participants also determined to a large extent who would benefit from the forest's wealth. Today's forest bears the traces of all the years of forest management that have come before. As policy changes, the new ideas are layered over the vestiges of past decisions—they cannot replace them. Each change in forest management both determines the practices of that time and attempts to correct the mistakes of the past. But not all mistakes can be easily corrected. In the Pacific Northwest today, trees grow according to reforestation protocols designed by five generations of foresters. Douglas fir trees can live for centuries, and some will always carry traces of how and why they were planted as they were.[23] Today's Douglas fir forest is a patchwork of history, reflecting in its patterns of growth all the past policies and beliefs of the region's foresters, lumber companies, and landowners. To manage Douglas fir forests today is to engage in two struggles. The first is the struggle to find harmony with those holding differing opinions of how that forest should be managed. The second is the struggle with generations of predecessors who practiced forestry in ways that sometimes fundamentally clashed with the science and values of the present day. Their decisions about planting, thinning, firefighting, and other aspects of management often remain all too evident in the forests of the present. The fight against these ghosts is often impossible to win, as the enemies have long since left the field of battle.

Let us revisit the Oregon and California Railroad revested lands, the former railroad grant lands discussed in chapter 5. Today, the discontinuous O&C lands still constitute a vast variety of Douglas fir forestlands in eighteen counties, spanning the state from the rugged canyon of the Rogue River to the well-traveled exurbia of the Willamette Valley. And that perplexing checkerboard of public land prompted the Department of the Interior to instigate the nation's first federal program of sustained-yield management in 1937. The long life of both trees and property lines creates tangible problems for people who live and work in the Douglas fir today, more than seventy-five years later. A checkerboard pattern still exists through much of this forest, delineating squares of public and private land. The public squares of the O&C lands are still controlled by the Department of the Interior, under the purview of the Bureau of Land Management (BLM), the successor to their original overseers, the General Land Office. The fates of the privately owned squares in between vary. Some are still managed for timber production, some have been developed for housing, some hold farms and orchards, and still others have been incorporated into the BLM holdings through land swaps. The forests on these public squares today vary widely in character, from intact forest remnants deep in the mountains to second- and third-growth industrial forest.[24]

While aiming to create an incentive for local communities to support sustained yield, the federal government unwittingly created a culture of dependence on the lumber industry. Woven into the original O&C legislation were stipulations that ensured that counties benefited more from O&C logging than from any other logging activity in the area. While counties normally receive 25 percent of gross receipts from logging on national forests within their borders, they receive 50 percent of gross receipts from logging on O&C lands. These "O&C Counties," mostly poor, sparsely populated, and remote, came to rely heavily on this revenue to provide infrastructure and services. For years the BLM authorized significantly more harvest from the O&C land in those counties than the Forest Service did from its lands.[25] As the counties' dependence on O&C revenue grew, the rate of logging on that land was often well above what sustained-yield calculations specified. As regulation tightened the parameters of logging in the region overall, the industry-friendly atmosphere and relatively lax rules of the BLM made O&C logging even more lucrative. Eventually, far in the future, the perennial overcut on the checkerboard might have finally forced a curtailment of the rate of logging. Before that day ever came, however, an unassuming little owl would overturn the entire scheme.

The recent fortunes of the northern spotted owl in the O&C lands illustrate well the clash between mainstream forestry's mode of industrially oriented management and the more recently developed framework of ecological management. The northern spotted owl (*Strix occidentalis caurina*) is medium sized, brown with white spots, nocturnal, and notably shy. It is one of three distinct spotted owl subspecies, which have a combined range spanning the forestlands of the North American West. The northern spotted owl is found in forest ecosystems of northern California, Oregon, Washington, and British Columbia. It relies largely on old-growth Douglas fir forest through most of its range, and on coast redwood forest in areas south of the Douglas fir region. It builds its nests in the cavities that form in the trunks of older living trees and in standing dead trees. As it requires a multilevel tree canopy for roosting and foraging, it can thrive only in a forest that retains most of its original array of young growth and secondary tree species. Tree Farms and highly managed forests have neither the multispecies composition nor the suitable nesting sites for this owl. The northern spotted owl was declared a threatened species in 1990 under the auspices of the Endangered Species Act. In 1994, ecologists estimated that 7.4 million acres of suitable, federally owned habitat were available for the owl throughout the American portions of its range, but safeguarding its survival would take more than simple acreage. The team of scientists brought together to devise a plan for the owl's survival found that it could not tolerate living near the edges of forestland, or in places where closed-canopy forest alternated with the open sky of recently logged areas. It did not thrive close to roads, logging, or any other human activity. This quirk of behavioral ecology meant scattered pockets of untouched forest would not suffice for the owl's habitat needs. Instead, large contiguous swaths of intact old growth were needed. In locations where the forest had already been fragmented too much, through logging or development, northern spotted owls were generally not found.[26]

When the federal government listed the northern spotted owl as a threatened species, it hit the O&C counties hard. The checkerboard layout of those forestlands had built fragmentation right into the design. There was almost no area of owl habitat in the O&C that was not already close to the minimum size that ecologists had determined the owl needed. That checkerboard, which had been intended to encourage industrial logging, now created the conditions that necessitated logging's complete cessation. While this area has long been the site of some of the Northwest's heaviest public-lands logging, the federally mandated spotted owl protection has stopped the O&C harvest entirely. Because of the local dependence on timber revenues for basic services, the shutdown hit not just those loggers and allied workers but all citizens of these counties. These remote and undeveloped towns had once spent beyond their means on infrastructure and services but now found themselves in dire straits with little hope of a return of the lost revenue. Antienvironmentalist sentiment ran high in these counties among loggers and nonloggers alike, as the full financial ramifications of the Northwest Forest Plan became apparent. The federal government was compelled to renege on the promises of a law grounded in industrial forestry in order to enforce a law grounded in ecological science. The new requirement that those timber sales cease in order to protect the owl superseded the earlier legal assurance that the counties and the lumber industry would benefit from O&C timber sales. The industrial forest management of the midcentury could not be harmonized with the ecological forest management that had been layered on top of it. Residents of these eighteen counties have been caught in the push and pull between the agendas of private industry and the federal government since the 1860s, when the would-be builders of the Oregon and California Railroad first took up the offer of checkerboarded land.[27]

Money Trees has described the struggles to possess—intellectually, governmentally, and economically—the Douglas fir and the forestlands it populates. From 1900 to 1944, the states of Washington and Oregon saw dramatic environmental and economic changes as their forests became places of encounter. This book has examined that era from the point of view of those professionals tasked with understanding and managing the Douglas fir as both a commodity and a living species. Over the course of those decades, the standard approaches to logging in Douglas fir country changed. Scientists also discovered more about the forests, the government acquired and defined lands, and communities throughout the region expanded. Through it all, the Douglas fir remained the totem species and economic focus—the money tree. Federal sustained yield may have changed the terms of the discussion immensely, but the disputes and drama have not abated. We can declare no real winners in this fight over Douglas fir, or indeed in any dispute over forest management. We often consider forest issues by adopting a prospective view, wondering whether a forest we love today will still exist for, still be loved by, the next generation. We would do well to include a retrospective view, too. A historical perspective helps illuminate the length and complexity of the project to understand and care for the forest. A constant flow of discourse among scientists and forest managers took place against the fluctuating demand for lumber and the ecological realities of the forest itself.

Notes

Abbreviations for Manuscript Collections

ALP   Aldo Leopold Papers
University of Wisconsin, Madison, WI

EFP   Emanuel Fritz Papers (BANC MSS 87/166)
Bancroft Library, University of California, Berkeley, CA

FDR   Franklin D. Roosevelt Manuscript Collections
Franklin D. Roosevelt Presidential Library, National Archives and
Records Administration, Hyde Park, NY

FRES   Records of the Forest and Range Experiment Station, Portland,
Oregon (MSS 95.10.4)
National Archives, Pacific Northwest and Alaska Region, Seattle, WA

HHCP   Herman Haupt Chapman Papers (MS 134)
Yale University, New Haven, CT

IFA   Industrial Forestry Association Records (MSS 1906)
Oregon Historical Society, Portland, OR

MMSFB   Natural Resources Department, Meeting Minutes of the State Forest
Board (AR 9-J-1)
Washington State Archives, Olympia, WA

NFPA   National Forest Products Association and National Lumber
Manufacturers Association Records (MSS 3395)
Forest History Society Library and Archives, Durham, NC

RMP   Robert Marshall Papers (BANC MSS 79/94)
Bancroft Library, University of California, Berkeley, CA

RZP   Raphael Zon Papers (P 1237)
Minnesota State Archives, St. Paul, MN

SAF   Society of American Foresters Records
Forest History Society Library and Archives, Durham, NC

SFC   Oregon State Forester's Correspondence
Oregon State Archives, Salem, OR

WCP   Weyerhaeuser Corporate Papers
Weyerhaeuser International Headquarters, Federal Way, WA

WCR   F. Weyerhaeuser and Company Records (P 983)
Minnesota State Archives, St. Paul, MN

WFP   Weyerhaeuser Family Papers (P 930)
Minnesota State Archives, St. Paul, MN

Introduction

1   Environmental historians argue that physical evidence of human interactions
     with landscapes, such as patterns of replanting, should be considered
     primary source material in environmental history. See William Cronon,
     "How to Read a Landscape," http://www.williamcronon.net/researching/
     landscapes.htm; William Cronon, "Kennecott Journey: The Paths Out of
     Town," in *Under an Open Sky: Rethinking America's Western Past,* ed.
     William Cronon, George Miles, and Jay Gitlin (New York: W. W. Norton,
     1992), 28–51; Tom Wessels, F*orest Forensics: A Field Guide to Reading the
     Landscape* (New York: W. W. Norton, 2010); and Tom Wessels, *Reading the
     Forested Landscape: A Natural History of New England* (Woodstock, VT:
     Countryman Press, 1999).

2   Aldo Leopold, *A Sand County Almanac* and *Sketches Here and There* (New
     York: Oxford University Press, 1949), 221.

3   Ibid.

4   For an illuminating dissection of Leopold's intentions and goals in writing
     Sand County Almanac, see Curt Meine, *Aldo Leopold: His Life and Work*
     (Madison: University of Wisconsin Press, 1988).

5   Henry Clepper provides a wide-angled view of the profession in *Professional
     Forestry in the United States* (Baltimore: Resources for the Future Press /
     The Johns Hopkins University Press, 1971). For the coalescence of a voice
     for wilderness preservation around resistance to federal development of
     wildlands, see Paul S. Sutter, *Driven Wild: How the Fight against Automobiles
     Launched the Modern Wilderness Movement* (Seattle: University of
     Washington Press, 2002). Sutter approaches the trained foresters Leopold
     and Marshall in the context of the wilderness movement, downplaying the
     importance of their careers in forestry. Char Miller considers Pinchot both
     during and after his time as chief forester in *Gifford Pinchot and the Making
     of Modern Environmentalism* (Washington, DC: Island Press, 2001). Aldo
     Leopold's criticism of forestry is discussed in Susan L. Flader, *Thinking like
     a Mountain: Aldo Leopold and the Evolution of an Ecological Attitude
     toward Deer, Wolves, and Forests* (Columbia: University of Missouri Press,
     1974), and in Meine, *Aldo Leopold.* For Bob Marshall, see James M. Glover,
     A Wilderness Original: The Life of Bob Marshall (Seattle: Mountaineers,
     1986); Roderick Nash, "The Strenuous Life of Bob Marshall," *Forest History
     Newsletter* 10, no. 3 (1966): 18–25; Sutter, *Driven Wild,* 194–238; and John
     P. Herron, *Science and the Social Good: Nature, Culture, and Community,
     1865–1965* (New York: Oxford University Press, 2010), 77–136.

6   A number of historians have examined the history of plant ecology, although
     fewer have worked on placing ecology in an environmental context.

Sharon E. Kingsland's *The Evolution of American Ecology, 1890–2000* (Baltimore: The Johns Hopkins University Press, 2005) and Peder Anker's *Imperial Ecology: Environmental Order in the British Empire, 1895–1945* (Cambridge, MA: Harvard University Press, 2002) have both ably depicted the interplay between field practice and theoretical sophistication and presented nuanced views of the social utility of ecology. Also important are Frank Golley, *A History of the Ecosystem Concept in Ecology: More Than the Sum of Its Parts* (New Haven, CT: Yale University Press, 1993); Ronald C. Tobey, *Saving the Prairies: The Life Cycle of the Founding School of American Plant Ecology, 1895–1955* (Berkeley: University of California Press, 1981); Donald Worster, *Nature's Economy: A History of Ecological Ideas* (Cambridge: Cambridge University Press, 1985); and Robert Kohler, *Landscapes and Labscapes: Exploring the Lab-Field Border in Ecology* (Chicago: University of Chicago Press, 2002).

7   Other historians who have examined forestry as a science include Nancy Langston, *Forest Dreams, Forest Nightmares: The Paradox of Old Growth in the Inland West* (Seattle: University of Washington Press, 1995); Richard Rajala, *Clearcutting the Pacific Rain Forest: Production, Science, and Regulation* (Vancouver: University of British Columbia Press, 1998); W. Scott Prudham, *Knock on Wood: Nature as Commodity in Douglas-Fir Country* (New York: Routledge, 2005); and Flader, *Thinking like a Mountain*. Stephen Bocking considers the role of scientists in many areas of environmental management in *Nature's Experts: Science, Politics, and the Environment* (New Brunswick, NJ: Rutgers University Press, 2004).

8   The classic, and still vital, history of federal land management in the Progressive Era is Samuel P. Hays, *Conservation and the Gospel of Efficiency: The Progressive Conservation Movement, 1890–1920* (1959; repr., Pittsburgh: University of Pittsburgh Press, 1999). Focusing specifically on forest management are Samuel P. Hays, *The American People and the National Forests* (Pittsburgh: University of Pittsburgh Press, 2009); and William G. Robbins, *Lumberjacks and Legislators: Political Economy of the U.S. Lumber Industry, 1890–1941* (College Station: Texas A&M University Press, 1982). For the US Northwest, see especially Langston, *Forest Dreams;* and David Louter, *Windshield Wilderness: Cars, Roads, and Nature in Washington's National Parks* (Seattle: University of Washington Press, 2006). For studies spanning US and Canadian management issues see Rajala, *Clearcutting;* and Prudham, *Knock on Wood.* With examinations restricted to the post–World War II era, important contributions are also made by Paul Hirt, *A Conspiracy of Optimism: Management of the National Forests since World War Two* (Lincoln: University of Nebraska Press, 1994); and Kevin

Marsh, *Drawing Lines in the Forest: Creating Wilderness Areas in the Pacific Northwest* (Seattle: University of Washington Press, 2010).

9   William G. Robbins also incorporated multiple levels of American forest management in *American Forestry: A History of National, State, and Private Cooperation* (Lincoln: University of Nebraska Press, 1985). Stephen Pyne discusses private-public cooperation of forest fire management in *Fire in America: A Cultural History of Wildland and Rural Fire* (Seattle: University of Washington Press, 1997). While focusing mainly on nineteenth-century settlers, Richard White elegantly incorporates multiple scales of forest management into his larger project in *Land Use, Environment, and Social Change: The Shaping of Island County, Washington* (Seattle: University of Washington Press, 1980).

Chapter 1

1   For the two men at the center of this story, see Michael P. Malone, *James J. Hill: Empire Builder* (Norman: University of Oklahoma Press, 1996); and Judith Koll Healey, *Frederick Weyerhaeuser and the American West* (St. Paul: Minnesota Historical Society Press, 2013). See also Ralph W. Hidy, Frank Ernest Hill, and Alan Nevins, *Timber and Men: The Weyerhaeuser Story* (New York: Macmillan, 1963); Robert E. Ficken, *The Forested Land: A History of Lumbering in Western Washington* (Seattle: University of Washington Press, 1987); and Charles E. Twining, *George S. Long: Timber Statesman* (Seattle: University of Washington Press, 1994).

2   Hidy, Hill, and Nevins, *Timber and Men*, 216; see also Twining, *George S. Long*.

3   The Pacific coast redwood of California is larger by some measures but does not grow as densely as the Douglas fir. The sequoia of the Sierra region can live longer but is not generally considered a timber tree. Jared Farmer, *Trees in California: A California History* (New York: W. W. Norton, 2013); Andrew C. Isenberg, *Mining California: An Ecological History* (New York: Hill and Wang, 2005).

4   For South Africa, see Hidy, Hill, and Nevins, *Timber and Men*, 219; for Hawaii, see Ficken, *Forested Land*; for Australia, see both Hidy, Hill, and Nevins, and Ficken. See also David Igler, *The Great Ocean: Pacific Worlds from Captain Cook to the Gold Rush* (New York: Oxford University Press, 2013).

5   Ficken, *Forested Land*, 25.

6   The Pacific coast redwood is the only commercial species exceeding Douglas fir in board feet per acre. Michael Williams, *Americans and Their Forests: A Historical Geography* (New York: Cambridge University Press, 1989).

7   As estimated by Ficken, *Forested Land*, 62.

8   Richard White and William Cronon have both considered the freight-pricing
    strategies of the railroads in *Railroaded: The Transcontinentals and the
    Making of Modern America* (New York: W. W. Norton, 2011) and *Nature's
    Metropolis: Chicago and the Great West* (New York: W. W. Norton, 1992),
    respectively. Neither focuses on the specific case at hand here. See also
    Ficken, *Forested Land*; Hidy, Hill, and Nevins, *Timber and Men*; Charles E.
    Twining, *F. K. Weyerhaeuser: A Biography* (St. Paul: Minnesota Historical
    Society Press, 1997); and Ralph W. Hidy, *The Great Northern Railway: A
    History* (Boston: Harvard Business School Press, 1988).

9   Between 1900 and 1914, members of the Weyerhaeuser family owned an
    interest in forty-eight separate logging and lumber companies in the upper
    Midwest, Southeast, northern Rocky Mountain, and Pacific Northwest
    sections of the nation. Furthermore, many of these forty-eight companies
    had multiple subsidiaries. For example, one of their largest in the Douglas
    fir region, the Weyerhaeuser Timber Company, held shares in twenty-seven
    different subsidiaries in 1914 (Hidy, Hill, and Nevins, *Timber and Men*).
    Because of the overlapping boards of directors of the many Weyerhaeuser
    companies and subsidiaries, individual family members and other executives
    were not usually affiliated with only a single company. I have identified the
    executives with the company or companies that are of most interest; other
    affiliations are not included.

10 The history of the Weyerhaeuser family and its business interests has been
    very thoroughly chronicled. An indispensable history to 1960 is Hidy, Hill,
    and Nevins, *Timber and Men*. For F. W. Weyerhaeuser, see Healey, *Frederick
    Weyerhaeuser*. See also Charles E. Twining, *Phil Weyerhaeuser, Lumberman*
    (Seattle: University of Washington Press, 1985); Charles E. Twining,
    A Bundle of Sticks: The Story of a Family (St. Paul: Rock Island, 1987);
    Charles E. Twining, *The Tie That Binds: A History of the Weyerhaeuser
    Family Office* (St. Paul: Rock Island, 1993); Twining, *George S. Long*; and
    Twining, *F. K. Weyerhaeuser*. See also the Weyerhaeuser-commissioned Joni
    Sensel, *Traditions through the Trees: Weyerhaeuser's First 100 Years* (Seattle:
    Documentary Book Publishers, 1999).

11 George S. Long to Frederick Weyerhaeuser Sr., January 20, 1903, 1 WFP.

12 George S. Long to Frederick Weyerhaeuser Sr., February 20, 1903, 1 WFP.

13 Twining, *George S. Long*; Ficken, *Forested Land*.

14 Williams, *Americans and Their Forests*; Clepper, *Professional Forestry*; Hirt,
    Conspiracy of Optimism.

15 Ernst Haeckel, *Natürliche Schöpfungsgeschichte: Gemeinverständliche
    Wissenschaftliche . . .* [The history of creation, or, the development of earth
    and its inhabitants . . .], trans. Ray Lankester (London: Henry S. King, 1876;
    Cambridge: Chadwyck-Healey, 1990); Eugenius Warming, *Plantesamfund:
    Grundtræk af den økologiske plantegeografi* [Oecology of plants: An

introduction to the study of plant communities], trans. Percy Groom and Isaac Bayley Balfour (Oxford: Clarendon Press, 1909; New York: Arno Press, 1977); Joel B. Hagen, *An Entangled Bank: The Origin of Ecosystem Ecology* (New Brunswick, NJ: Rutgers University Press, 1992); Sharon E. Kingsland, *Modeling Nature: Episodes in the History of Population Ecology* (Chicago: University of Chicago Press, 1995); Worster, *Nature's Economy*; Tobey, *Saving the Prairies*.

16   Henry Chandler Cowles, "The Ecological Relations of the Vegetation on the Sand Dunes of Lake Michigan. Part I: Geographical Relations of the Dune Floras," *Botanical Gazette* 27, no. 2 (1899): 96.

17   Ibid., 112.

18   Charles Darwin, *On the Origin of Species by Means of Natural Selection, or the Preservation of Favoured Races in the Struggle for Life*, 1st ed. (London: John Murray, 1865; facsimile repr. Cambridge, MA: Harvard University Press, 1964), 74–75.

19   Philip J. Pauly, *Biologists and the Promise of American Life: From Meriwether Lewis to Alfred Kinsey* (Princeton, NJ: Princeton University Press, 2000); Richard G. Beidleman, *California's Frontier Naturalists* (Berkeley: University of California Press, 2006); Andrew Menard, *Sight Unseen: How Frémont's First Expedition Changed the American Landscape* (Lincoln: University of Nebraska Press, 2012).

20   Tobey, *Saving the Prairies*.

21   Robert P. McIntosh, *The Background of Ecology: Concept and Theory* (Cambridge: Cambridge University Press, 1991); Golley, *History of the Ecosystem Concept*; Hagen, *Entangled Bank*. For a critique of autecology, see Eugene P. Odum, *Fundamentals of Ecology* (Philadelphia: W. B. Saunders, 1953); and Betty Jean Craige, *Eugene Odum: Ecosystem Ecologist and Environmentalist* (Athens: University of Georgia Press, 2001).

22   Gifford Pinchot, *A Primer of Forestry, Part I: The Forest*, USDA Division of Forestry Bulletin, no. 24, vol. 1 (1899), 44.

23   Ibid., 45.

24   Ibid., 64.

25   Gregg Mitman, *The State of Nature: Ecology, Community, and American Social Thought, 1900–1950* (Chicago: University of Chicago Press, 1992); Kingsland, *Evolution of American Ecology*; McIntosh, *Background of Ecology*; Golley, *History of the Ecosystem Concept*; Hagen, *Entangled Bank*.

26   Gifford Pinchot, *Breaking New Ground* (New York: Harcourt, Brace, 1947; repr., New York: Island Press, 1998), 255. Citations refer to the Island Press edition.

27   Ibid.

28   Andrew Denny Rodgers II, *Bernhard Eduard Fernow: A Story of North American Forestry* (Princeton, NJ: Princeton University Press, 1951); Carl A. Schenck, *Cradle of Forestry in America: The Biltmore Forest School, 1898–1913*, ed. Ovid Butler (Durham, NC: Forest History Society, 1998); Miller, *Gifford Pinchot*; Pinchot, *Breaking New Ground*; Hidy, Hill, and Nevins, *Timber and Men*.

29   John Muir, "The Hetch Hetchy Valley," *Sierra Club Bulletin* 6, no. 4 (1908): 211–20, subsequently revised and reprinted in *The Yosemite* (New York: Century, 1912).

30   T. C. C. [Thomas Chrowder Chamberlin], "Editorial," *Journal of Geology* 18, no. 5 (1910): 468–69. A quote extracted from the original editorial also appears in Hays, *Conservation and the Gospel of Efficiency*, 262–63. Hays cites an unsigned editorial condemning Chamberlin: "A Lost Leader," *American Forests* 16, no. 10 (1910): 604–6.

31   "What Is Conservation?" *New York Times*, August 18, 1910.

32   "Lost Leader," 606.

Chapter 2

1   Robert Righter, "National Monuments to National Parks: The Use of the Antiquities Act of 1906," *Western Historical Quarterly* 20 (August 1989): 281–301; Ben W. Twight, *Organizational Values and Political Power: The Forest Service and the Olympic National Park* (University Park: Pennsylvania State University Press, 1983).

2   Jim Kjeldsen, *The Mountaineers: A History* (Seattle: Mountaineers Books, 1998).

3   Twight, *Organizational Values and Political Power*; "Washington: Mount Olympus Park," *Time*, July 11, 1938.

4   While the Forest Service was actively trapping and killing predators in the 1930s, the National Park Service's stated aim was preserving the complete array of species. However, there was also ambivalence, even resistance, to national park designation from many wilderness advocates. Many declared the Park Service too focused on visitor comfort and access, pointing to the network of roads constructed in previously pristine areas of many national parks. Automobiles allowed easy access into a park, obliterating that very remoteness and ruggedness that had been so important to the original aesthetic. The threat of overdomesticating the wildness of the Olympics concerned many in the Mountaineers, who were nonetheless some of the strongest backers of park designation. Alternatively, maintaining the area as a national forest or national monument could limit touristic development, but the attendant threat of logging and industrial development would increase. In a January 26, 1935, letter to the Mountaineers' Irving Clark,

Bob Marshall went so far as to declare, "I do not feel that it would be any safer as a National Park than as a National Forest" (Sutter, *Driven Wild*, 234). The concerns of the Mountaineers and the Wilderness Society were not unfounded. The National Park Service's master plan declared Olympic National Park a "wilderness park" and highlighted both the scarcity of roads and its intention to preserve that scarcity, although one of the first Park Service projects was the creation of a sightseeing road for tourists (Louter, *Windshield Wilderness*).

5   See Righter, "National Monuments to National Parks," for a discussion of the malleability of the Antiquities Act.

6   Despite his importance and his intriguing biography, Zon has often been neglected in histories of ecology and environment, but see Char Miller, "Militant Forester, Raphael Zon," *Forest Magazine*, Winter 2005, 22–23; Norman J. Schmaltz, "Forest Researcher: Raphael Zon, Part I," *Journal of Forest History* 24, no. 1 (1980): 25–39; Norman J. Schmaltz, "Forest Researcher: Raphael Zon, Part II," *Journal of Forest History* 24, no. 2 (1980): 86–97; Paul O. Rudolph, "R. Zon, Pioneer in Forest Research," *Science* 125, no. 3261 (1957): 1283–84; Jeremy Young, "Roots of Research: Raphael Zon and the Origins of Forest Experiment Stations," in *Fort Valley Experimental Forest: A Century of Research 1908–2008*, ed. Susan D. Olberding and Margaret M. Moore, USDA Proceedings of the Rocky Mountain Research Station P-55 (2008), 363–70, reprinted in edited form as Jeremy Young, "Warrior of Science: Raphael Zon and the Origins of Forest Experiment Stations," *Forest History Today*, Spring/Fall 2010, 4–12. A worthwhile, if brief, chronicle of Zon's early scholarship in Europe and New York can be found in Rodgers, *Fernow*, 339–40.

7   The father of Alexander Kerensky, seven years younger than Zon, was the director of the secondary school that Kerensky, Ulyanov, and Zon all attended. Richard Abraham, *Alexander Kerensky: The First Love of the Revolution* (New York: Columbia University Press, 1987).

8   David E. Lewis, "The University of Kazan: Provincial Cradle of Russian Organic Chemistry," *Journal of Chemical Education* 71, no. 1 (1994): 93–97.

9   Schmaltz, "Raphael Zon, Part I"; Schmaltz, "Raphael Zon, Part II"; Rudolph, "R. Zon." Zon studied within the faculty of Medicine at Kazan, which had only four general faculties: Physics/Mathematics, History/Philosophy, Law, and Medicine. Lewis, "University of Kazan."

10   Robert J. Richards, *The Tragic Sense of Life: Ernst Haeckel and the Struggle over Evolutionary Thought* (Chicago: University of Chicago Press, 2008); Frederick B. Churchill, "Weismann, Hydromedusae, and the Biogenetic Imperative: A Reconsideration," in *A History of Embryology*, ed. T. J. Horder, J. A. Witkowski, and C. C. Wylie (Cambridge: Cambridge University

Press, 1985): 7–33; Garland E. Allen, "T. H. Morgan and the Split between Embryology and Genetics, 1910–1935," in Horder, Witkowski, and Wylie, *History of Embryology*, 113–146.

11   C. B. Metz, "The Naples Zoological Station and the Marine Biological Laboratory: One Hundred Years of Biology," supplement, *Biological Bulletin* 168 (1985): 1–207.

12   The town of Tilsit is now known as Sovetsk, in the Russian exclave of Kaliningrad Oblast between Lithuania and Germany.

13   Raf de Bont, "Evolutionary Morphology in Belgium: The Fortunes of the 'Van Beneden School,' 1870–1900," *Journal of the History of Biology* 41, no. 1 (2008): 81–118.

14   Rodgers, *Fernow*, 339.

15   Ibid.; Schmaltz, "Raphael Zon, Part II."

16   Mark Bevir, *The Making of British Socialism* (Princeton, NJ: Princeton University Press, 2011); G. D. H. Cole, *British Working Class Politics, 1832–1914* (London: G. Routledge and Sons, 1941).

17   George Bernard Shaw, *The Fabian Society: Its Early History*, Fabian Tract no. 41 (London: Fabian Society, 1892), 5.

18   Ibid., 16.

19   Edward R. Pease, *History of the Fabian Society* (New York: E. P. Dutton, 1916), 50; Schmaltz, "Raphael Zon, Part I."

20   Anker, *Imperial Ecology*, 11.

21   Schmaltz, "Raphael Zon, Part I"; Schmaltz, "Raphael Zon, Part II"; Rudolph, "R. Zon"; Miller, "Militant Forester"; Young, "Roots of Research."

22   Rodgers, *Fernow*, 339.

23   Zon to Samuel P. Dana, August 8, 1944, quoted in Schmaltz, "Raphael Zon, Part II," 96. See also Miller, "Militant Forester."

24   Ralph A. Hosmer, "Education in Professional Forestry," in *Fifty Years of Forestry in the U.S.A.*, ed. Robert K. Winters (Washington, DC: Society of American Foresters, 1950), 299–315; Clepper, *Professional Forestry*. For German influence see also Flader, *Thinking like a Mountain*; Meine, *Aldo Leopold*; and Aldo Leopold, "Deer and Dauerwald in Germany: I. History," *Journal of Forestry* 34, no. 4 (April 1936): 366–75.

25   In 1903, the state of New York closed down the Cornell school by canceling the entire appropriations bill that funded it, although this was more because of larger arguments among state politicians than because of Fernow's leadership. Gifford Pinchot consistently criticized Fernow for his decisions on the School of Forestry's curriculum and governance. For Pinchot's allegations, see Pinchot, *Breaking New Ground*, 151, 184. See also Hosmer, "Education in Professional Forestry"; and Rodgers, *Fernow*.

26   For the criticisms, see Pinchot, *Breaking New Ground*; and Miller, *Gifford Pinchot*.

27 Williams, *Americans and Their Forests*; Clepper, *Professional Forestry*; Robbins, *American Forestry*; Harold K. Steen, *Forest Service Research: Finding Answers to Conservation's Questions* (Durham, NC: Forest History Society, 1998); Margaret Herring and Sarah Greene, *Forest of Time: A Century of Science at Wind River Experimental Forest* (Corvallis: Oregon State University Press, 2007).

28 Clepper, *Professional Forestry*; Schmaltz, "Raphael Zon, Part II"; Miller, "Militant Forester"; Rudolph, "R. Zon."

29 Thornton T. Munger described arriving at the Wind River Experimental Forest in 1908: "People looked down on research in those days. . . . In general the more romantic administrative frontier work appealed to them more than what they considered the ring-counting work of a researcher." Herring and Greene, *Forest of Time*, 24.

30 The passage in question in Matthew's work concerns mainly the detrimental effects of inbreeding among European hereditary nobles, but it begins with general observations. "There is a law universal in nature, tending to render every reproductive being the best possibly suited to its condition that its kind, or that organized matter, is susceptible of, which appears intended to model the physical and mental or instinctive powers, to their highest perfection, and to continue them so. This law sustains the lion in his strength, the hare in her swiftness, and the fox in his wiles. As Nature, in all her modifications of life, has a power of increase far beyond what is needed to supply the place of what falls by Time's decay, those individuals who possess not the requisite strength, swiftness, hardihood, or cunning, fall prematurely without reproducing." Patrick Matthew, *On Naval Timber and Arboriculture* (London: Longman, Rees, Orme, Brown, and Green, 1831; electronic facsimile ed., http://darwin-online.org.uk/content/frameset?viewtype=side&itemID=A154&pageseq=1), 364–65.

31 Raphael Zon, "Darwinism in Forestry," *American Naturalist* 47, no. 561 (September 1913): 540. Zon cites Francis Darwin, *The Life and Letters of Charles Darwin* (New York: Appleton, 1898), 95.

32 Zon, "Darwinism in Forestry," 541. Zon cites the "Historical Sketch" in the 1878 edition of *Origin of Species*, xvi. Note: Zon's citation is to the American edition; this version of the Historical Sketch first appeared in the 1876 English edition of *Origin* and is not found in all editions. Janet Browne, *Darwin's Origin of Species: A Biography* (New York: Atlantic Monthly Press, 2006).

33 Zon, "Darwinism in Forestry," 544.

34 Ibid., 546.

35 Julian Huxley, introduction to *Animal Ecology*, by Charles S. Elton (London: Sidgwick and Jackson, 1927), xiv–xv.

36 Edgar Nelson Transeau, "The Accumulation of Energy by Plants." *Ohio Journal of Science* 26, no. 1 (1926): 1–10. See also Paul A. Colinvaux, "The Efficiency of Life," in *Why Big Fierce Animals Are Rare: An Ecologist's Perspective* (Princeton, NJ: Princeton University Press, 1978), 32–46.

37 Terminology of Charles O. Whitman; see Pauly, *Biologists*, 162. Whitman's passenger pigeon episode also appears in Mark Barrow, *Nature's Ghosts: Confronting Extinction from the Age of Jefferson to the Age of Ecology* (Chicago: University of Chicago Press, 2009).

38 Andrew Isenberg, *The Destruction of the Bison: An Environmental History, 1750–1920* (New York: Cambridge University Press, 2000).

39 John Muir, "Our National Forests," *Atlantic Monthly* 80, no. 478 (August 1897): 146.

40 Elton, *Animal Ecology*, 105–7.

41 Ibid., 106.

42 Anker, *Imperial Ecology*.

43 Young, "Roots of Research."

44 Some recent historical scholarship has asserted that experimental fieldwork was a strange new concept to a generation of biologists trained in the lab, and that the maturation of American ecology required pushing past this to develop viable methodologies. However, when the discipline of forestry is included within the realm of biology, this interpretation is less tenable. Kohler, *Landscapes and Labscapes*; Jeremy Vetter, "Rocky Mountain High Science: Teaching, Research and Nature at Field Stations," in *Knowing Global Environments: New Historical Perspectives on the Field Sciences*, ed. Jeremy Vetter (New Brunswick, NJ: Rutgers University Press, 2011), 108–34.

45 Pauly, *Biologists*.

46 Patricia Craig, *Centennial History of the Carnegie Institution of Washington*, vol. 4, *The Department of Plant Biology* (New York: Cambridge University Press, 2005); Thomas D. Cornell, "Carnegie Institution of Washington," in *History of Science in the United States: An Encyclopedia*, ed. Marc Rothenberg (New York: Garland, 2001), 102–8; Robert E. Kohler, *Partners in Science: Foundations and Natural Scientists, 1900–1945* (Chicago: University of Chicago Press, 1991); James D. Ebert, "Carnegie Institution of Washington and Marine Biology: Naples, Woods Hole, and Tortugas," *Biological Bulletin* 168 (1985): 172–82.

47 Janice E. Bowers, *A Sense of Place: The Life and Work of Forrest Shreve* (Tucson: University of Arizona Press, 1988); Tad Nichols and Bill Broyles, "Afield with Desert Scientists," *Journal of the Southwest* 39 (Autumn-Winter 1997): 353–70; Robert P. McIntosh, "Pioneer Support for Ecology," *BioScience* 33 (February 1983): 107–12; Frederick Vernon Coville and Daniel Trembly MacDougal, *Desert Botanical Laboratory of the Carnegie*

Institution (Washington, DC: Carnegie Institution, 1903). See also Kohler, *Landscapes and Labscapes*; and Craig, *Centennial History.*

48   William F. Ganong, "The Cardinal Principles of Ecology," *Science* 19, no. 482 (March 25, 1904): 493–98. See also Eugene Cittadino, *Nature as the Laboratory: Darwinian Plant Ecology in the German Empire, 1880–1900* (New York: Cambridge University Press, 1990).

49   Ganong, "Cardinal Principles," 493.

50   Ibid., 494.

51   Jeremy Vetter's survey of fieldwork at the research stations of the US Rocky Mountains neglects the Fremont Station but addresses the work of the Alpine Lab. Vetter, "Rocky Mountain High Science." See also Edith S. Clements, *Adventures in Ecology* (New York: Pageant Press, 1960); and Tobey, *Saving the Prairies.*

52   Rodgers, *Fernow*, 410; Clements, *Adventures in Ecology.*

53   For Fort Valley, see Olberding and Moore, *Fort Valley Experimental Forest.*

54   On the collaborator program, see Pinchot, *Breaking New Ground*, 149–50; US Department of Agriculture, *A National Plan for American Forestry* [Copeland Report]. S. Doc. No. 12, 73rd Cong., 1st sess., March 13, 1933; Clepper, *Professional Forestry*; Robbins, *American Forestry.*

55   Frederic E. Clements, *Life History of Lodgepole Burn Forests*, USDA Forest Service Bulletin 79 (Washington, DC: Government Printing Office, 1910), 56.

56   Quoted in a letter (n.d.) to Fernow in Rodgers, *Fernow*, 411. Clements's lodgepole study is also discussed briefly in Kohler, *Landscapes and Labscapes*, 228–29.

57   Zon's work in subsequent years would critique Clementsian ecological practice in more depth, as well. See especially the widely read Carlos G. Bates and Raphael Zon, *Research Methods in the Study of Forest Environment*, Bulletin of the US Department of Agriculture, no. 1059 (Washington, DC: Government Printing Office, 1922).

58   Ibid.

59   Raphael Zon and Henry S. Graves, *Light in Relation to Tree Growth*, Bulletin of the US Department of Agriculture, no. 93 (1911), 59.

60   Rodgers, *Fernow*, 421.

61   Ibid.

62   Henry C. Cowles, "An Ecological Aspect of the Conception of Species," *American Naturalist* 42, no. 496 (1908): 265.

63   For discussions of instrument design, choice, and procurement, see Bates and Zon, *Research Methods.*

64   Richard S. Sartz, "Carlos G. Bates: Maverick Forest Service Scientist," *Journal of Forest History* 21, no. 1 (1977): 31–39; Bates and Zon, *Research Methods*; George G. Ice and John D. Stednick, "Forest Watershed Research in the United States," *Forest History Today*, Spring/Fall 2004, 16–26.

65   McIntosh, *Background of Ecology*; Golley, *History of the Ecosystem Concept*. For Clements's ecological climax see also Tobey, *Saving the Prairies*; and Kingsland, *Evolution of American Ecology*.

66   Kingsland makes a similar argument in her discussion of Gleason, though she does not relate it to forestry. *Evolution of American Ecology*, 161.

67   Watershed studies including assessments of drinking water, commercial fishing, and agriculture go back to Carlos Bates's study in 1909. Such inclusive management can be seen as the precursor to the "Working Circle" mode of federal forest management seen in the post–World War II era. Ice and Stednick, "Forest Watershed Research."

Chapter 3

1   Two oral histories describe the Wind River Experiment Station and the earliest years of research there: Thornton T. Munger and Amelia R. Fry, *Forest Research in the Northwest* (Berkeley: Regional Oral History Office, Bancroft Library, University of California, 1967); and Leo A. Isaac and Amelia R. Fry, *Douglas Fir Research in the Pacific Northwest, 1920–1956* (Berkeley: Regional Oral History Office, Bancroft Library, University of California, 1967). See also Herring and Greene, *Forest of Time*; and Ivan Doig, *Early Forestry Research: A History of the Pacific Northwest Forest and Range Experiment Station, 1925–1975* (Washington, DC: USDA Forest Service, 1976).

2   Emily K. Brock, "The Challenge of Reforestation: Ecological Experiments in the Douglas Fir Forest, 1920–1940," *Environmental History* 9, no. 1 (2004): 57–79; Rajala, *Clearcutting*, 95–96; Leo A. Isaac and Amelia R. Fry, "The Seed-Flight Experiment: Policy Heeds Research," *Forest History* 16 (October 1973): 54–60; J. V. Hofmann, "Natural Reproduction from Seed Stored in the Forest Floor," *Journal of Agricultural Research* 11 (October 1917): 1–26.

3   Hofmann, "Natural Reproduction," 3.

4   Ibid., 12.

5   Ibid.; J. V. Hofmann, "The Establishment of a Douglas Fir Forest," *Ecology* 1 (1920): 49–53; J. V. Hofmann, "Furred Forest Planters," *Scientific Monthly* 16 (1923): 280–83; Brock, "Challenge of Reforestation."

6   Hofmann, "Natural Reproduction," 20.

7   First-year results in Hofmann, "Natural Reproduction," 21; failure of subsequent years in Isaac and Fry, "Seed-Flight Experiment."

8   Hofmann, "Natural Reproduction," 20n1–2. For an example of a wide review of tree-seed germination studies see M. Büsgen, *The Structure and Life of Forest Trees*, 3rd ed., revised by E. Münch, trans. Thomas Thomson (London: Chapman and Hall, 1929), 392–98.

9   Isaac and Fry, *Douglas Fir Research*, 65.

10   Hofmann, "Natural Reproduction," 7.

11   Hofmann, "Establishment," 53.

12   Rajala, *Clearcutting*, 111–19; Isaac and Fry, "Seed-Flight Experiment"; Munger and Fry, *Forest Research*; Brock, "Challenge of Reforestation."

13   Isaac to Marshall, June 2, 1927, 1/7/20 RMP.

14   Isaac to Marshall, May 12, 1937, 1/7/20 RMP.

15   Leo A. Isaac, "Seed Flight in the Douglas Fir Region," *Journal of Forestry* 28 (1930): 492–99.

16   Leo A. Isaac, "Life of Douglas Fir Seed in the Forest Floor," *Journal of Forestry* 33 (1935): 61–66.

17   Foresters had difficulty determining success and failure of reforestations. Because the experimental projects were centered on only one species, only through that species could the experiment be understood. Douglas fir grow very slowly; a decade can pass before the full results of a single planting begin to appear. When confronted with one failure, preventing a host of others could be impossible. In 1945 the American Forestry Association produced a survey of the forest conditions of the Douglas fir region. Out of 26 million acres of commercial forestland, reforestation following cutting or fire had failed on 3.3 million acres. However, all of that land had been deforested prior to 1930. No conclusion could yet be drawn about the condition of the 2.1 million acres of land deforested between 1930 and 1945. The report noted that "much of this land was logged under systematic methods designed to ensure reseeding," yet the author would still declare neither success nor failure. The slow-moving bureaucracy of the Forest Service and the obstinacy of the lumber industry only amplified this problem. J. B. Woods, "Report of the Forest Resource Appraisal," *American Forests* 52: 413–28. Isaac noted that Hofmann's seed-storage theory still had its adherents well into the 1960s. Isaac and Fry, *Douglas Fir Research*, 69–70.

18   Isaac to Marshall, June 2, 1927, 1/7/20 RMP. The two would periodically exchange friendly correspondence on personal, professional, and scientific matters until Marshall's death in 1939.

19   Among other positions, Louis Marshall was the founding member of the American Jewish Committee, chairman of the board of directors of the Jewish Theological Seminary of America, and the president of the prominent Congregation Emanu-El of New York City. Robert Marshall, *Bob Marshall in the Adirondacks: Writings of a Pioneering Peak-Bagger, Pond-Hopper, and Wilderness Preservationist*, ed. Phil Brown (Saranac Lake, NY: Lost Pond Press, 2006); Glover, *Wilderness Original*; Herron, *Science and the Social Good*; M. M. Silver, *Louis Marshall and the Rise of Jewish Ethnicity in America: A Biography* (Syracuse: Syracuse University Press, 2013).

20   Hosmer, "Education in Professional Forestry," 299–315.

21   Marshall, *Bob Marshall in the Adirondacks*; Glover, *Wilderness Original*.

22   This college paper was republished, with some revisions, as Robert Marshall, "Recreational Limitations of Silviculture in the Adirondacks," *Journal of Forestry* 23, no. 2 (1925): 173–78.

23   Isaac and Fry, *Douglas Fir Research*, 56–61.

24   Robert Marshall, "Precipitation and Presidents," *The Nation* 124 (March 23, 1927): 316. While Marshall began "Precipitation and Presidents" with an anecdote from his student days at Syracuse, his data were from weather stations as far back as 1850.

25   Ellsworth Huntington, "The Secret of the Big Trees," *Harper's Magazine* 125 (July 1912): 292–302. Interestingly, in the decades since Marshall wrote "Precipitation and Presidents," it has occasionally been cited as a serious piece of research by scholars in a variety of social science disciplines. Neil J. Smelser, *Theory of Collective Behavior* (London: Routledge and Kegan Paul, 1962), 55n3; Stephen J. Bahr and Boyd C. Rollins, "Crisis and Conjugal Power," *Journal of Marriage and the Family* 33, no. 2 (May 1971): 360–67; Christian Ben Lakhdar and Eric Dubois, "Climate and Electoral Turnout in France," *French Politics* 4, no. 2 (2006): 137–57. The meteorological magazine *Weatherwise* featured a reprint of Marshall's entire article accompanied by an editorial comment that "the drought of 1988 will undoubtedly play some role in determining whether George Bush or Michael Dukakis is our next president," proving, perhaps, that success in comedy consists in finding the right audience. *Weatherwise* 41, no. 5 (October 1988): 263–65.

26   Robert Marshall, "Influence on Precipitation Cycles in Forestry," *Journal of Forestry* 25, no. 4 (1927): 415–29.

27   Isaac and Fry, *Douglas Fir Research*, 56–61. Other similar science parodies by Marshall included a study of the average number of pancakes eaten by his Forest Service superiors while visiting the regional Forest Service station, described in Glover, *Wilderness Original*, and in Robert Marshall, "Contribution to the Life History of the Northwestern Lumberjack," *Social Forces* 8 (December 1929): 271–73.

28   Robert Weidman to Marshall, July 16, 1926, quoted in Glover, *Wilderness Original*, 84.

29   US Department of Agriculture, *National Plan for American Forestry*, 667–68; Glover, *Wilderness Original*, 83–84.

30   The prominent Frederic E. Clements, among others, had considered plant physiology a part of ecology, stating that "ecology bears the closest of relations [to physiology, and] . . . it is, indeed, nothing but a rational field physiology, which regards form and function as inseparable phenomena," and that ecology was "normalizing [physiology] by forcing it into the field

as the place for experiment, and by directing the chief attention to the plant as an organism rather than a complex of organs, is also rapidly coming to be felt." Frederic E. Clements, *Research Methods in Ecology* (Lincoln, NE: University Publishing Company, 1905), 17.

31 Because his most widespread contributions came not in the form of published papers but through the development of innovative technical methods and the manufacture of highly specialized field instruments, his impact on the field of plant ecology has been somewhat understated by historians. See Kohler, *Landscapes and Labscapes*; and Kingsland, *Evolution of American Ecology*.

32 Term from Kingsland, *Evolution of American Ecology*.

33 Among other concerns, the water content of plants was found to fluctuate significantly over the course of a single day, and abiotic factors influenced water content as well, making the estimates of limited value. Robert Marshall, "An Experimental Study of the Water Relations of Seedling Conifers with Special Reference to Wilting" (PhD diss., Johns Hopkins University, 1930), 42.

34 Ibid., 41.

35 Riichiro Koketsu, another Livingston student, had used the mimosa plant, in which it was "possible to observe ocularly the beginning of temporary wilting and the onset of some subsequent stages." Among hardier species such as conifers and cacti, however, even in the youngest seedlings "it is almost or quite impossible to decide by direct observation just when any stage of wilting has begun." Ibid., 42.

36 Livingston and Forrest Shreve had done similar work with cacti at the Carnegie Desert Lab, fitting the plants with field sensors to take measurements in situ. Bowers, *Sense of Place*; see also Kohler, *Landscapes and Labscapes*.

37 Marshall, "Experimental Study of Water Relations," 43.

38 Bates and Zon, *Research Methods*. Livingston's estimation of that publication was tepid, mainly because of the difficult transitions between sophisticated discussion of advances in ecological technique and simplistic discussion of the mechanics of Forest Service protocol. However, Livingston praised Bates and Zon for their discussion, particularly of the complex question of water absorption and soil moisture in tree roots. Burton E. Livingston, "Research Methods in Ecology," *Ecology* 5, no. 1 (January 1924): 99–101. This review's title is the same as that of Clements's widely known publication, *Research Methods in Ecology*. It is unclear to what extent the implication that the title under review replaces that famous text was intended by either Livingston or the journal's editors.

39 Livingston had been aware of, and actively engaged with, the application of quantitative methods of measurement in forest ecology well prior to Marshall's arrival in his lab. Livingston, "Research Methods."

40   Western yellow pine and western white pine. See Marshall, "Experimental Study of Water Relations," 44, table 1.

41   Slash pine and shortleaf pine. Ibid.

42   White spruce and "Northern white cedar," better known as the ornamental American arborvitae. Ibid.

43   The diversity and utility of test species allowed Marshall to generate findings useful to federal foresters. While all coniferous, they came from six different genera with differing physiological and ecological characteristics. If Marshall had designed his inquiry to maximize physiological knowledge, the species list might have consisted of an array of *Picea* species, rather than species as disparate as Norway pine and western red cedar. Particularly the inclusion of the lone *Abies*, which is ecologically important but has such soft wood that it is of essentially no commercial use, and the two *Thuja*, which were used mainly for ornamentals and fence posts, speaks to Marshall's agenda as an ecologist first, rather than a forester or physiologist.

44   Marshall, "Experimental Study of Water Relations," 88.

45   Ibid.

46   Ibid., 91. Marshall often abstracted or obscured the scientific content of his work for an audience that he assumed had no interest or training in the sciences. Historians have tended to accept Marshall's public black-boxing of his scientific work as evidence of its biographical unimportance. However, such an approach has left dark both Marshall's intellectual development and the forging of his important allegiances within ecology. In his letters to family and other nonscientists, Marshall wrote of the tedium of city life and did not discuss his research. Glover, in his 1986 biography of Marshall, interpreted that along with certain passages in Marshall's writings about Alaska as a lack of true commitment to his stated scientific interests (*Wilderness Original*, 104–5). However, those passages are far more ambiguous than Glover asserts and must be read alongside his evinced commitment to ecological research in other writings. Marshall's younger brother George Marshall, in an introduction to *Arctic Wilderness*, a posthumous collection of his older brother's Alaska diaries and articles, makes a claim similar to Glover's, but with similarly weak evidence. Robert Marshall, *Arctic Wilderness*, edited and with an introduction by George Marshall (Berkeley: University of California Press, 1956). Marshall's words, his actions, and his choice of research topic show that he had never intended to leave federal forestry behind for an academic career. Glover also suggested that Marshall's immediate return to the administrative life was more evidence that his time in Baltimore was merely a flight of fancy in a life otherwise dedicated to wilderness. However, that interpretation does not correlate with Marshall's continual support of science, and particularly ecology, throughout his life. Marshall's ecological

education must be understood as a quirky, yet genuine, manifestation of his commitment to scientific inquiry, a testament to his broad intelligence, and a demonstration of his genuine curiosity about the natural world.

47   Exceptions were the northern white cedar, which came from the New York State Conservation Commission, and the shortleaf pine, which was sourced from the Louisiana Department of Conservation. Marshall, "Experimental Study of Water Relations."

48   Ibid.

49   During the first period of American control in the Philippines the islands were administered by the US Army, and hence American military officers were appointed to many of the managerial roles. Ahern had lectured on forestry during his time stationed in Montana in the 1880s, where he had also befriended Pinchot, who in turn recommended him for the Philippines post. He never worked in the Forest Service and did not have a forestry degree, although his thesis at Yale Law School had been on forest legislation. Pinchot, *Breaking New Ground*; Clepper, *Professional Forestry*; Gerald W. Williams, *The Forest Service: Fighting for Public Lands* (Westport, CT: Greenwood Publishing Group, 2007); Richard P. Tucker, "Unsustainable Yields: American Foresters and Tropical Timber Resources," in *Insatiable Appetite: The United States and the Ecological Degradation of the Tropical World* (Berkeley: University of California Press, 2000), 345–415.

50   The mention of the *Nation* article is a reference to Marshall, "Precipitation and Presidents."

51   Marshall to Ahern, [n.d., January 1929, handwritten copy of the letter sent], 1/5/3 RMP.

52   Glover, *Wilderness Original*; Schmaltz, "Raphael Zon, Part I"; Schmaltz, "Raphael Zon, Part II"; Clepper, *Professional Forestry*.

53   Later, to the mainstream audiences that read his *Arctic Village*, he was modest in his reference to the scientific content of his work this first summer. Not wishing to dwell on technicalities that would bore nonspecialists, he wrote that in planning the first visit he determined that "there ought to be some purpose back of this spree, so I rationalized a scientific investigation as a reason for my expedition. As a forester and a plant physiologist, it seemed eminently appropriate that I should make a study of tree growth at northern timberline." Robert Marshall, *Arctic Village* (New York: Literary Guild, 1933), 3.

54   Gifford Pinchot, *To the South Seas: The Cruise of the Schooner Mary Pinchot to the Galapagos, the Marquesas, and the Tuamotu Islands, and Tahiti* (Philadelphia: John C. Winston, 1930), 5. Pinchot's own trip was no doubt inspired in part by Theodore Roosevelt's Smithsonian-Roosevelt African Expedition (1909–1910), during which the newly retired Roosevelt, assisted by three Smithsonian Museum researchers, collected specimen

animals as they traveled through British East Africa, the Belgian Congo, and Sudan. Theodore Roosevelt, *African Game Trails: An Account of the African Wanderings of an American Hunter-Naturalist* (New York: Charles Scribner's Sons, 1910); J. Lee Thompson, *Theodore Roosevelt Abroad: Nature, Empire, and the Journey of an American President* (New York: Palgrave Macmillan, 2010).

55 Marshall, *Arctic Wilderness*, xlii; Martin Wilmking and Jens Ibendorf, "An Early Tree-Line Experiment by a Wilderness Advocate: Bob Marshall's Legacy in the Brooks Range, Alaska," *Arctic* 57, no. 1 (2004): 106–13. Marshall's hypothesis derives from Gleasonian ecological theories. Henry A. Gleason, "The Vegetational History of the Middle West," *Annals of the Association of American Geographers* 12 (1922): 39–85.

56 Marshall was also tangentially involved in the early stages of Harold Ickes's attempt to establish a community of displaced Jews in Alaska under the auspices of the Department of the Interior. Marshall died before the effort fell apart in failure with the publication of the Slattery Report of 1940. For Marshall's involvement see "Committee on Refugee Problems, Report on Alaska 1939," 3/3/47 RMP. See also Gerald S. Berman, "Reaction to the Resettlement of World War II Refugees in Alaska," *Jewish Social Studies* 44, no. 3/4 (1982): 271–82.

57 Marshall to F. H. Eyre, April 12, 1939, 1/10/19 RMP.

58 Marshall, *Arctic Village*; Marshall, *Alaska Wilderness*; Glover, *Wilderness Original*.

59 Marshall describes conducting timber cruises several times in his writings on Alaska but does not use the technical term "timber cruise." See, for example, Marshall, *Alaska Wilderness*, 53.

60 Ibid., 60.

61 Marshall to Al Cline, July 15, 1930, 1/6/2 RMP. See also "The Alatna and the John," in Marshall, *Alaska Wilderness*, 82–109.

62 Marshall, *Arctic Village*, 3.

63 Marshall to Gerry and Lily Kempff, March 3, 1930, 1/8/4 RMP.

64 Shepard to Marshall, July 1, 1930, 1/9/11 RMP.

65 Ibid.

66 Pinchot, *To the South Seas*, 5.

67 Likewise, Leopold's detailed journal investigations of the behavior and ecology around his weekend shack fit into this category. Leopold, *Sand County Almanac*; Meine, *Aldo Leopold*.

68 Marshall, *Arctic Village*, 3.

69 Ibid., 9. *Arctic Village*, for all Marshall's attempts to create an objective portrait of the entire town, excels most in its humorous and touching descriptions of the tragedies, celebrations, philosophy, and hard-won life lessons of individual Alaskans. The book's dedication reads, "To the people

of the Koyukuk who have made for themselves the happiest civilization of which I have knowledge" (v).

70 Robert Marshall, "Adventure, Arrogance, and the Arctic," *Wings*, May 1933, 9–12, republished in partial form in Marshall, *Alaska Wilderness*, 1–4.

71 Christopher M. Granger to Zon, January 14, 1933, 7 RZP.

72 Marshall to Gerry and Lily Kempff, March 3, 1930, 1/8/4 RMP.

73 George P. Ahern, Robert Marshall, E. N. Munns, Gifford Pinchot, Ward Shepard, W. N. Sparhawk, and Raphael Zon, "A Letter to Foresters," *Journal of Forestry* 28 (April 1930): 456–58. See also Douglas Brinkley, *The Quiet World: Saving Alaska's Wilderness Kingdom, 1879–1960* (New York: HarperCollins, 2011); Sutter, *Driven Wild*; Rajala, *Clearcutting*; and Herron, *Science and the Social Good*.

74 Ahern et al., "Letter to Foresters," 456.

75 Ibid.

76 Ibid., 457 (all quotations in this paragraph).

77 Ibid., 456. This follows the approach Shepard had recommended in 1927, when he declared that in order to establish effective measures of public control "we as foresters must have full faith in forestry . . . and must approach our task with the covenanter's proselyting zeal. Our cause is a moral cause of the first magnitude, because it affects the weal and happiness of millions of people through many generations." While the letter makes use of religious language and allusion throughout, it is important for us to recognize that it was not appealing to a specifically Christian view of the role of humans as caretakers of the earth. Indeed, among the list of seven signatories were the Jewish Marshall and Zon, both of whom were widely recognized as members of that faith among the Society of American Foresters. Ward Shepard, "The Necessity for Realism in Forestry Propaganda," *Journal of Forestry* 25 (January 1927): 14.

78 R. S. Kellogg, "As I See It," *Journal of Forestry* 28 (April 1930): 462.

79 Ward Shepard, a Forest Service forester who became the American representative of the German American Carl Schurz Memorial Foundation in 1933, was in 1930 living in Washington, DC, yet no longer listing an affiliation with the Forest Service or any other entity. See Ward Shepard, "Cooperative Control: A Proposed Solution of the Forest Problem," *Journal of Forestry* 28 (February 1930): 113–20.

80 Ahern et al., "Letter to Foresters," 463.

81 A similar problem occurs with the word "conservation." For industrial claims of "doing forestry," see Emily K. Brock, "Tree Farms on Display: Presenting Industrial Forests to the Public in the Pacific Northwest, 1941–1960," *Oregon Historical Quarterly* 113 (Winter 2012): 526–59. See also Hidy, Hill, and Nevins, *Timber and Men*; and Clepper, *Professional Forestry*.

82   Robert Marshall, "The Problem of the Wilderness," *Scientific Monthly* 30, no. 2 (February 1930): 141–48.

83   Ibid., 145. See also William Cronon, "The Trouble with Wilderness; or, Getting Back to the Wrong Nature," in *Uncommon Ground: Rethinking the Human Place in Nature*, ed. William Cronon (New York: W. W. Norton, 1995), 69–90; Sutter, *Driven Wild*; and Herron, *Science and the Social Good*.

84   Stuart to Marshall, March 7, 1930, 3.2/3/30 RMP.

85   Stuart to district foresters of all districts, March 11, 1929, 3.2/3 RMP.

86   Ibid.

87   Memorandum, "Amendment of National Forest Manual," July 22, 1929, 3.2/3 RMP.

88   Clapp to Marshall, November 3, 1930, 3.2/3/35 RMP.

89   Ibid.

90   The establishment of the collaborator designation is also discussed in chapter 2. The ranks of the collaborators included a number of esteemed academics, including Nathaniel E. Shaler and George E. Bessey, as well as Frederic Clements.

91   Henry A. Wallace, "Letter of Transmittal," in US Department of Agriculture, *National Plan for American Forestry*, vii.

92   Ibid., viii.

93   "The Copeland Report," *American Forests*, May 1933, 211.

94   Marshall, "The Forest for Recreation," in US Department of Agriculture, *National Plan for American Forestry*, 469.

95   Aldo Leopold would call it "monumental" upon its publication, although that comment reflected more admiration for its exhaustiveness and wide distribution than agreement with its conclusions. Aldo Leopold, "Conservation Economics," *Journal of Forestry* 32, no. 5 (1934): 537.

96   US Department of Agriculture, *National Plan for American Forestry*, 654.

97   Ibid.

98   Ibid., 661.

99   Robert Marshall, *The People's Forests* (New York: H. Smith and R. Haas, 1933; Iowa City: University of Iowa Press, 2002), acknowledgments.

100 Robert Marshall, *The Social Management of American Forests* (New York: League for Industrial Democracy, 1930).

101 Marshall, *People's Forests*.

102 Ibid., dedication.

103 Glover, *Wilderness Original*; Sutter, *Driven Wild*.

104 Leopold, "Conservation Economics," 537.

Chapter 4

1   For the story of the initial 1933 fires see Pyne, *Fire in America*. See also
    Ellis Lucia, *Tillamook Burn Country* (Cadwell, ID: Caxton Printers, 1983);
    William G. Robbins, *Landscapes of Promise: The Oregon Story, 1800–1940*
    (Seattle: University of Washington Press, 1997); Gail Wells, *The Tillamook:
    A Created Forest Comes of Age* (Corvallis: Oregon State University Press,
    1999); and Stewart Holbrook, *Burning an Empire: The Study of American
    Forest Fires* (New York: Macmillan, 1943).

2   Richard M. Highsmith Jr. and John L. Beh, "Tillamook Burn: The
    Regeneration of a Forest," *Scientific Monthly* 75, no. 3 (September 1952):
    139–48.

3   Ibid.; Wells, *Created Forest*.

4   On New Deal and the environment, see among other sources Gary Dean
    Best, *Pride, Prejudice, and Politics: Roosevelt versus Recovery* (New York:
    Praeger, 1991); William J. Barber, *Design within Disorder: Franklin D.
    Roosevelt, the Economists, and the Shaping of American Economic Policy*
    (Cambridge: Cambridge University Press, 1996); Matthew J. Dickinson,
    *Bitter Harvest: F. D. R., Presidential Power, and the Growth of the
    Presidential Branch* (Cambridge: Cambridge University Press, 1997); Clara
    Juncker, "Bureaucratic Dynamics and Control of the New Deal's Publicity:
    Struggles between Core and Periphery in the FSA's Information Division," in
    The Roosevelt Years: New Perspectives on American History, 1933–1945,
    ed. Robert Garson and Stuart S. Kidd (Edinburgh: Edinburgh University
    Press, 1999); A. L. Reisch Owen, *Conservation under FDR* (New York:
    Praeger, 1983); Irving Brant, *Adventures in Conservation with Franklin
    D. Roosevelt* (Flagstaff, AZ: Northland Publishing, 1988); and Karen R.
    Merrill, *Public Lands and Political Meaning: Ranchers, the Government,
    and the Property between Them* (Berkeley: University of California Press,
    2002).

5   Franklin D. Roosevelt, *Franklin D. Roosevelt and Conservation: 1911–
    1945*, comp. and ed. Edgar B. Nixon (Hyde Park, NY: Franklin D. Roosevelt
    Library, 1957), 130. See also Clepper, *Professional Forestry*.

6   Marshall to Zon, January 23, 1933. Robert Marshall Collection, 1 FDR.

7   The National Recovery Administration was deemed unconstitutional in a
    Supreme Court decision, Schecter Poultry Corporation v. United States, in
    May 1935, but the impact of the NRA on industry lingered for some time.

8   Mark Reed to the Pacific Northwest Loggers Association, June or July 1933.
    Quoted in Ficken, *Forested Land*, 196.

9   F. E. Weyerhaeuser to J. P. Weyerhaeuser Jr., July 12, 1933. Quoted in Ficken,
    Forested Land, 197.

10 For John B. Woods as a Long-Bell forester and a lumber industry champion,
    see Robbins, *American Forestry*; on the Lumber Code, see John B.

Woods, "Status of the Article X Joint Conservation Program," *Journal of Forestry* 33, no. 12 (1935): 958–63; John B. Woods, "The Post-Code Status of Conservation," *Journal of Forestry* 33, no. 8 (1935): 710–12; H. H. Chapman, "Second Conference on the Lumber Code," *Journal of Forestry* 32, no. 3 (1934): 272–307; and A. B. Recknagel, "Summary of Forest Practice Rules under the Conservation Code (Article X)," *Journal of Forestry* 33, no. 9 (1935): 792–95.

11 Neil M. Maher, *Nature's New Deal: The Civilian Conservation Corps and the Roots of the American Environmental Movement* (New York: Oxford University Press, 2009); John A. Salmond, *The Civilian Conservation Corps, 1933–1942: A New Deal Case Study* (Durham, NC: Duke University Press, 1967).

12 Besides the Lumber Code and the CCC, various other New Deal programs also affected the community of professional foresters. As head of the Department of Agriculture, Henry Wallace, with his rethinking of supply and demand for farm products, would influence the parallel problem of stabilizing the lumber industry. Arthur M. Schlesinger Jr., *The Age of Roosevelt: The Coming of the New Deal* (Boston: Houghton Mifflin, 1958). See also Norman D. Markowitz, *The Rise and Fall of the People's Century: Henry A. Wallace and American Liberalism, 1941–1948* (New York: Free Press, 1973); John C. Culver and John Hyde, *American Dreamer: The Life and Times of Henry A. Wallace* (New York: W. W. Norton, 2000); and Donald Worster, *The Dust Bowl: The Southern Plains in the 1930s* (New York: Oxford University Press, 1979).

13 Raphael Zon, "Forests and Water in the Light of Scientific Investigation," in *Final Report of the National Waterways Commission*, S. Doc. No. 469, 62nd Cong., 2nd sess. (1912), appendix 5; Raphael Zon, "Forests as a Factor in Flood Control within the Upper Mississippi River Valley," in *Relation of Forestry to the Control of Floods in the Mississippi Valley*, H.R. Doc. No. 573, 70th Cong., 2nd sess. (1929), 85–113; E. N. Munns, "Watershed and Related Forest Influences," in US Department of Agriculture, *National Plan for American Forestry*, 299–462.

14 Owen, *Conservation under FDR*; Schlesinger, *Age of Roosevelt*.

15 Schmaltz, "Raphael Zon, Part II"; Owen, *Conservation under FDR*; Winters, *Fifty Years of Forestry*.

16 Marshall to Ickes, August 25, 1935, 1/7/17 RMP.

17 Zon Petition, June 13, 1934, 1a SAF. The Zon Petition was subsequently printed in the October issue of the *Journal*. George P. Ahern, Carlos G. Bates, Earle H. Clapp, L. F. Kneipp, W. C. Lowdermilk, Robert Marshall, E. N. Munns, Gifford Pinchot, Edward C. M. Richards, F. A. Silcox, William N. Sparhawk, and Raphael Zon, "The Petition of June 13, 1934," *Journal of Forestry* 32, no. 7 (1934): 781–83.

18 See chapters 2 and 3 of this book for more discussion of Bates as Zon's coauthor.

19 Raphael Zon and William N. Sparhawk, *Forest Resources of the World* (New York: McGraw-Hill, 1923); Raphael Zon and William N. Sparhawk, *America and the World's Woodpile*, Circular of the US Department of Agriculture, no. 21 (1928).

20 Ward Shepard and Franz Heske, "European Facts for American Skeptics," *Journal of Forestry* 31, no. 8 (1933): 923–31; Arthur R. Hogue, "The Carl Schurz Memorial Foundation: The First Twenty-Five Years," *Indiana Magazine of History* 51, no. 4 (1955): 335–39.

21 Zon to Pinchot, May 23, 1933, 6 RZP.

22 Ahern et al., "Petition of June 13," 782.

23 Ibid.

24 Shirley W. Allen, "The Society of American Foresters," in Winters, *Fifty Years of Forestry*, 272–84.

25 The version of the Society of American Foresters' constitution in effect in 1934 called for an editorial staff made up of eight society members chosen by the president of the society, plus an editor in chief elected by the society's Executive Council. Society of American Foresters' Constitution of 1928, 83/832 HHCP.

26 H. H. Chapman, "Making the Society of American Foresters a Professional Organization," *Journal of Forestry* 32, no. 4 (1934): 503–7.

27 A. Forester [pseud.], "A Research Forester Looks at His Job," *Journal of Forestry* 32, no. 5 (1934): 598–603.

28 Emanuel Fritz, "Report of the Editor: Resignation Tendered," *Journal of Forestry* 31, no. 2 (1933): 233–35. Emphasis in original.

29 Reed was originally asked to stay on as editor just until the annual meeting in December 1933. At that time he was asked to extend his double duties until May 1934. In fact, Reed would remain listed on the *Journal* masthead as the acting editor in chief until January 1935.

30 Clepper, *Professional Forestry*. Reed and Chapman's complaints about the lack of New Deal–related article submissions was voiced in Franklin Reed, "An Editorial Policy for the Journal of Forestry," *Journal of Forestry* 32, no. 7 (1934): 785–91.

31 Ahern et al., "Petition of June 13," 782.

32 Memorandum, Reed to Chapman, July 14, 1934, 47/ 144 HHCP.

33 For example, after receiving the draft of Marshall's response to criticism of the petition, Reed wrote to Chapman that "he now definitely puts yourself and [Ward] Shepard in the holy of holies class who are beyond reproach and narrows his attack to Fritz and myself as being the arch and dirty villains in the plot." Reed to Chapman, October 3, 1934, 1a SAF. The Marshall response would be published in the November issue of the *Journal*.

34 H. H. Chapman, "Statement by President Chapman," *Journal of Forestry* 32 (October 1934): 777–81.
35 Reed to Chapman, January 8, 1934 [sic], 1a SAF. From the context, this letter was clearly written in January 1935, not 1934 as stated.
36 Chapman to Reed, January 10, 1935, 1a SAF.
37 David Montgomery, "Evolution of the Society of American Foresters 1934–1937 as Seen in the Memoirs of H. H. Chapman," *Forest History Newsletter* 6, no. 3 (1962): 2–9.
38 H. H. Chapman, "The Responsibilities of the Profession in the Present Situation," *Journal of Forestry* 33, no. 3 (1935): 204–14; "Proceedings, 34th Annual Meeting of the Society of American Foresters," *Journal of Forestry* 33, no. 3 (1935): 189–90.
39 Munns to Fritz, October 10, 1934, 1a SAF, 1.
40 Kneipp to Fritz, November 30, 1934, 1a SAF.
41 Wackerman to Reed, November 1934, 1a SAF.
42 "Editorial Petition of June 13, 1934: Executive Secretary Reed's Letter of July 18," 26 SAF.
43 J. O. Hazard query letter response, n.d., 26/8-5 SAF. Hazard, a political appointee, was a fervent supporter of Roosevelt programs and ideals; see J. O. "Hap" Hazard, "The Roosevelt Forestry Song," *American Forests* 39 (October 1933): 250.
44 H. M. Meloney query letter response, July 27, 1934, 26/8-5 SAF.
45 For example, R. C. Hawley, an associate editor and Yale silviculture professor, wrote, "Under no circumstances allow the journal to be controlled by any one small group. Do not be afraid to tell such a group to start its own publication." R. C. Hawley query letter response, July 24, 1934, 26/8-5 SAF.
46 Julian E. Rothery, a New York–based forest engineer with the International Paper Company, wrote that "there is certainly no objection to the publication of another organ which will reflect the policy of the New Deal and perhaps it would be helpful, but it should be plainly known as such an organ and stand on its own feet as such." Julian E. Rothery query letter response, September 8, 1934, 26/8-5 SAF.
47 R. S. Kellogg query letter response, July 20, 1934, 26/8-5 SAF.
48 Hawley query letter response, n.d., 26/8-5 SAF.
49 S. R. Black query letter response, July 24, 1934, 26/8-5 SAF.
50 Charles W. Boyce query letter response, August 2, 1934, 26/8-5 SAF.
51 C. L. Van Giesen query letter response, July 26, 1934, 26/8-5 SAF.
52 S. J. Hall query letter response, July 20, 1934, 26/8-5 SAF.
53 Clepper, *Professional Forestry*; Robbins, *American Forestry*.
54 Burt P. Kirkland query letter response, August 10, 1934, 26/8-5 SAF.
55 Reed to Fritz [n.d.], 1/1 EFP.
56 [H. H. Chapman], "Editorial: Professional Idealism," *Journal of Forestry* 32, no. 7 (1934): 677–79.

57 "Society Affairs," *Journal of Forestry* 32, no. 7 (1934): 777–96.

58 Emanuel Fritz, "Statement by Former Editor-in-Chief Emanuel Fritz," *Journal of Forestry* 32, no. 7 (1934): 785.

59 Ahern et al., "Petition of June 13," 781.

60 Reed to Chapman, August 16, 1934, 1a SAF, 2.

61 Ibid. Punctuation as in original.

62 Ibid.

63 Montgomery, "Evolution of the Society." See also *Journal of Forestry*, 1933–1934.

64 Herron, Science and the Social Good; Glover, *Wilderness Original*.

65 Thomas W. Alexander to H. H. Chapman, January 21, 1935, 1a SAF, 3. Emphasis in original.

66 Some combined their dislike for Franklin Roosevelt with xenophobia to conclude that New Deal forestry was not American enough. Fletcher went on, "Coming as I do from a family who first settled the first inland town in Mass. (Concord 1631) my background is one that can't stand for the Socialistic ideas now being forced on the Country." E. D. Fletcher to H. H. Chapman, September 22, 1934, 1a SAF.

67 Front matter, *Journal of Forestry* 33, no. 1 (1935): 1.

68 Ibid.

69 Pamphlet to foresters, 1 SAF. Thomas W. Alexander to H. H. Chapman, January 21, 1935, 1a SAF.

70 Robert Marshall, "Fallacies in Osborne's Position," *Living Wilderness* 1 (1935): 4; Robert Marshall, "Comments on Commission's Truck Trail Policy," *American Forests* 43 (January 1936): 6, 48. Sutter, *Driven Wild*; Glover, *Wilderness Original*; Meine, *Aldo Leopold*.

71 [Herbert A. Smith], "Editorial: The Cult of the Wilderness," *Journal of Forestry* 33 (December 1935): 955–57.

72 Ibid., 957.

73 Sutter, *Driven Wild*; Meine, *Aldo Leopold*; Glover, *Wilderness Original*.

74 Leopold to Zon, May 19, 1926, 8/001/001/63-64 ALP.

75 Aldo Leopold query letter response, July 21, 1934, 26/8-5 SAF.

76 Meine, *Aldo Leopold*; Sutter, *Driven Wild*; Flader, *Thinking Like a Mountain*; Roderick Nash, *Wilderness and the American Mind* (New Haven, CT: Yale University Press, 1967).

Chapter 5

1 Elwood R. Maunder, *A Forester's Log: Fifty Years in the Pacific Northwest; An Interview with George L. Drake* (Durham, NC: Forest History Society, 2004).

2 Hays, *Conservation and the Gospel of Efficiency*.

3   See Ficken, *Forested Land*. That issue was not specific to the Northwest;
     see Williams, *Americans and Their Forests*; and Andrew Gennett, *Sound
     Wormy: Memoir of Andrew Gennett, Lumberman* (Athens: University of
     Georgia Press, 2002).

4   Weyerhaeuser sometimes attributed its corporate commitment to lumber
     to its founders' Germanic traditions of family and forest. European forests
     were not a resource that could be used and then abandoned. Ownership
     of forestland was considered a permanent investment to be tended well.
     Since the seventeenth century, English and Continental foresters had been
     concerned with the exhaustion of forest resources. Reforestation strategies
     after logging were considered of utmost importance in German forests.
     The forestry practiced at the Clemons Tree Farm was quite different from
     that of the German forest plantations of the nineteenth and early twentieth
     centuries, but the belief that forests might be renewed through artificial
     reforestation and intensive forest management are long standing. Healey,
     Frederick Weyerhaeuser; Hidy, Hill, and Nevins, *Timber and Men*; Ficken,
     Forested Land; Twining, *Phil Weyerhaeuser*.

5   "The Weyerhaeuser Idea as to Reforestation," *American Forests* 16 (March
     1910): 194. While there were a number of different companies under the
     Weyerhaeuser umbrella, they usually functioned under the same corporate
     philosophy and policies. The statements of a spokesperson for one
     Weyerhaeuser-affiliated company usually applied to all the companies and
     their subsidiaries.

6   "Weyerhaeuser Idea as to Reforestation," 194; Twining, *George S. Long*.

7   Twining, *George S. Long*; Robbins, *American Forestry*; Rajala, *Clearcutting*.

8   Twining, *George S. Long*.

9   Roderic Olzendam, *Green Gold for America: The Life and Times of Roderic
     Marble Olzendam* (Portland, OR: Binford and Mort, 1982); William B.
     Greeley, *Forests and Men* (New York: Doubleday, 1951); "Weyerhaeuser
     Idea as to Reforestation."

10   [Raphael Zon?], "What Is Industrial Forestry?" *Journal of Forestry* 26, no.
     1 (1928): 2. Emphasis in original.

11   Brock, "Challenge of Reforestation."

12   Chapman to Fritz, November 2, 1931, 3/S3/12 EFP.

13   Ibid.; see also L. F. Kneipp, "Industrial Forestry," *Journal of Forestry* 26, no.
     4 (1928): 507–12.

14   Kneipp, "Industrial Forestry."

15   Ibid., 508.

16   Elmo Richardson, *David T. Mason, Forestry Advocate: His Role in the
     Application of Sustained Yield Management to Private and Public Forest
     Lands* (Durham, NC: Forest History Society, 1983).

17   Chapter 13, "The Law of Supply and Demand," in Ficken, *Forested Land*, describes the regional lumber economics of this era well. See also Williams, *Americans and Their Forests*; and Robbins, *Landscapes of Promise*. Langston discusses a similar case in another forest type in *Forest Dreams*.

18   Emily K. Brock, "New Patterns in Old Places: Forest History for the Global Present," in *The Oxford Handbook of Environmental History*, ed. Andrew C. Isenberg (New York: Oxford University Press, 2014), 154–77; Michael Williams, *Deforesting the Earth: From Prehistory to Global Crisis* (Chicago: University of Chicago Press, 2002).

19   Hays, *American People*.

20   David T. Mason to John W. Blodgett Sr., April 2, 1928, 166/ 31 WFP, 1. See also Historical Note, John W. Blodgett Jr. papers, Coll. 223, Special Collections and University Archives, University of Oregon Libraries, Eugene, OR; Richardson, *David T. Mason*.

21   Mason to Blodgett, April 2, 1928, 1–2.

22   Quoted from the journal *West Coast Lumberman*, in Ficken, *Forested Land*, 177.

23   Greeley, *Forests and Men*; Clepper, *Professional Forestry*.

24   Carl M. Stevens, Donald Bruce, and Harvey C. Jack, "Clearwater Timber Company," n.d., 168 WCR. See also Carl M. Stevens to Minot Davis, February 19, 1930, 168 WCR; and Twining, *Phil Weyerhaeuser*.

25   Donald Bruce, one of Mason's partners, recommended "that an immediate request should be made to the Forest Service to investigate the possibilities of a joint sustained yield operation involving your holdings and those of the Clearwater National Forest." Donald Bruce to C. L. Billings, October 18, 1929, 168 WCR.

26   Carl M. Stevens to Evan W. Kelley, June 17, 1930, 168 WCR.

27   Donald Bruce to H. H. Chapman, October 16, 1930, 168 WCR; Twining, *Phil Weyerhaeuser*.

28   In the plan's first multiyear cutting cycle, the prescription was to cut 50 percent of the lucrative white pine on the Clearwater Timber Company land, while sparing all the less valuable stands of larch and fir. The projected second cutting cycle depended on one of two things happening: either the Forest Service would grant the company inexpensive contracts to harvest the white pine on adjacent public land, or a rise in market price for the still-standing less valuable larch and fir would make its harvest economically justified. The consultants predicted that the price for larch and fir would rise over the course of the next two decades as a result of widely anticipated advances in Weyerhaeuser mills' processed wood products. Donald Bruce to H. H. Chapman, October 16, 1930, 168 WCR; Hidy, Hill, and Nevins, *Timber and Men*.

29   Donald Bruce to H. H. Chapman, October 16, 1930, 168 WCR.

30   Richardson, David T. Mason; Twining, *Phil Weyerhaeuser*. The next attempted cooperative effort was not until 1946, with the Shelton Cooperative Sustained Unit in Shelton, WA. See David A. Clary, "What Price Sustained Yield? The Forest Service, Community Stability, and Timber Monopoly under the 1944 Sustained-Yield Act," *Journal of Forest History* 31, no. 1 (January 1987): 4–18.

31   David T. Mason, *Forests for the Future: The Story of Sustained Yield as Told in the Diaries and Papers of David T. Mason, 1907–1950*, ed. Rodney C. Loeher (St. Paul: Forest Products History Foundation / Minnesota Historical Society, 1950); Clepper, *Professional Forestry*; Hidy, Hill, and Nevins, *Timber and Men*.

32   Elmo Richardson, *BLM's Billion Dollar Checkerboard: Managing the O&C Lands* (Santa Cruz, CA: Forest History Society, 1980); Derrick Jensen and George Draffan, *Railroads and Clearcuts: Legacy of Congress's 1864 Northern Pacific Railroad Land Grant* (Spokane, WA: Inland Empire Public Lands Council, 1995); Robbins, *Landscapes of Promise*. For railroad bankruptcy more generally, see White, *Railroaded*; and David A. Skeel, "Railroad Receivership and the Elite Reorganization Bar," in *Debt's Dominion: A History of Bankruptcy Law in America* (Princeton, NJ: Princeton University Press, 2001), 48–72.

33   Richardson, *David T. Mason*; Mason, *Forests for the Future*.

34   John P. Weyerhaeuser Jr. to Frederick E. Weyerhaeuser. Quoted in Twining, *Phil Weyerhaeuser*, 148. Emphasis in original.

35   Frederick E. Weyerhaeuser was the president of the Tacoma-based Weyerhaeuser Timber Company from 1934 to 1945, the period discussed in this chapter. However, F. E. Weyerhaeuser, still based in Saint Paul, Minnesota, had little time to spend on the daily intricacies of the Weyerhaeuser Timber Company's Northwest operations. F. E. Weyerhaeuser was, at the time, also the president of the Weyerhaeuser Sales Company and was on the board of directors of numerous other Weyerhaeuser subsidiaries. For an exhaustive historical overview of the officers of the major lumber and logging Weyerhaeuser subsidiaries, see "Appendix IX: Officers and Directors of Five Companies," in Hidy, Hill, and Nevins, *Timber and Men*, 658–65.

36   John P. Weyerhaeuser Jr. to Frederick E. Weyerhaeuser, January 1936. Quoted in Twining, *Phil Weyerhaeuser*, 160.

37   J. P. Weyerhaeuser Jr. to F. E. Weyerhaeuser, 1937. Quoted in Twining, *Phil Weyerhaeuser*, 192. See also Hidy, Hill, and Nevins, *Timber and Men*.

38   Hidy, Hill, and Nevins, *Timber and Men*; Twining, *Phil Weyerhaeuser*.

39   Olzendam to J. P. Weyerhaeuser Jr., February 1937. Quoted in Twining, *Phil Weyerhaeuser*, 180. Olzendam had previously been employed as the director of the Metropolitan Life Insurance Company's Social Security Bureau and in that role had developed effective strategies for mediating corporate, governmental, and public needs. Olzendam, *Green Gold*.

40   George P. Ahern, *Deforested America: Statement of the Present Forest Situation in the United States* (Washington, DC: privately printed, 1928); Marshall, *People's Forests*; Susan R. Schrepfer, *The Fight to Save the Redwoods: A History of Environmental Reform, 1917–1978* (Madison: University of Wisconsin Press, 1983); Hays, *Conservation and the Gospel of Efficiency.*

41   Franklin Roosevelt to the 75th session of the US Congress, April 29, 1938, quoted in Stuart Ewen, *P.R.: A Social History of Spin* (New York: Basic Books, 1996), 293. For the cultural context of advertising and public relations in America, see T. J. Jackson Lears, "Part II: The Containment of Carnival: Advertising and American Social Values from the Patent Medicine Era to the Consolidation of Corporate Power," in *Fables of Abundance: A Cultural History of Advertising in America* (New York: Basic Books, 1994).

42   Ewen, *P.R.*; Larry Tye, *The Father of Spin: Edward L. Bernays and the Birth of Public Relations* (New York: Crown Publishing, 1998).

43   Advertisement, "What Is Your America All About?" in Ewen, *P.R.*, 304. See also Tye, *Father of Spin.*

44   Roderic Olzendam, "Timber Is a Crop: An Address Delivered at the Pacific Logging Congress, Seaside, Oregon," Weyerhaeuser Timber Company in-house publication, 1937, 7/Public Affairs WCP. Capitalization as in the original.

45   Ibid., 9.

46   Ibid., 15.

47   Ibid., 16, emphasis in original.

48   Ibid., 17.

49   Ibid., 18.

50   Twining, *George S. Long.* See also Twining, *Phil Weyerhaeuser*, 183.

51   William B. Greeley, "Economic Aspects of Our Timber Supply," *Journal of Forestry* 20, no. 8 (1922): 837–47.

52   Thornton T. Munger, "Forest Research in the Ryderwood Region," [n.d., circa 1926], File 2.01 FRES, 3. See also A. H. Hodgson, [untitled memorandum], July 9, 1926, File 2.11 FRES; A. H. Hodgson, "Report on the Forestry Activities of the Long Bell Lumber Company, Longview, Washington," July 9, 1926, File 2.01 FRES.

53   William B. Greeley, "Industrial Forestry," in Winters, *Fifty Years of Forestry*, 238–59; William B. Greeley, Earle H. Clapp, Joseph Kittredge, Ward Shepard, Herbert A. Smith, William N. Sparhawk, and Raphael Zon, "Timber: Mine or Crop?" in *US Department of Agriculture Yearbook 1922* (Washington, DC: Government Printing Office, 1923), 83–180.

54   The idea of timber as a crop had been prevalent for centuries in Europe. In the eighteenth and nineteenth centuries, many foresters had written about timber as being, at least metaphorically, an agricultural crop. The earliest use of this metaphor found by this historian occurred in 1726. The English

botanist Richard Bradley weighed the relative benefits of planting vegetables, shrub fruit, or timber trees on a site, addressing timber as one among several possible cash crops. He wrote, "I have heard, that herbs grow in pence, and shrubs in shillings, while timber grows in pounds." Richard Bradley, *General Treatise of Husbandry and Gardening, Containing a New System of Vegetation* (London: T. Woodward, 1726). See also Brock, "New Patterns in Old Places."

55   Gifford Pinchot, *A Primer of Forestry, Part II: Practical Forestry*, USDA Division of Forestry Bulletin, no. 24, vol. 2 (1905), 12.

56   Weyerhaeuser Sales Company's advertisements and marketing since 1928 had focused on its premium brand of lumber, named "4-SQUARE." The marketing of the brand-stamped 4-SQUARE lumber had been based on the innovation of trimming the lumber to specified lengths with right-angled ends. This innovation was coupled with refinements in packaging and distribution that improved the appearance of the lumber in the lumberyard, increasing sales and inspiring a host of imitations. Hidy, Hill, and Nevins, *Timber and Men*.

57   Hidy, Hill, and Nevins, *Timber and Men*; Twining, *Phil Weyerhaeuser*.

58   See Brock, "Tree Farms on Display."

59   Roderic Olzendam, "'We the People' and the Clemons Tree Farm," [remarks at the June 12 dedication of the Clemons Tree Farm], 3/Speeches/Olzendam WCP, 2.

60   On wilderness rhetoric, see Nash, *Wilderness and the American Mind*; James Morton Turner, *The Promise of Wilderness: American Environmental Politics since 1964* (Seattle: University of Washington Press, 2012); Sutter, *Driven Wild*; and Cronon, "The Trouble with Wilderness."

61   Chapman seemed to admit the government's case had merit: "If possible, the presentation, if there is one, *and there likely will be* [emphasis added], should be as logical and orderly as possible, so that a good case and record may be built up." C. S. Chapman to J. P. Weyerhaeuser Jr., "Proposed Congressional Committee to Investigate Forest Practices of Industry," [interoffice memorandum], March 28, 1938, 3/Correspondence/J. P. Weyerhaeuser Jr. 1938 WCP, 1.

62   J. P. Weyerhaeuser Jr. to C. S. Chapman, April 7, 1938, 3/Correspondence/J. P. Weyerhaeuser Jr. 1938 WCP.

63   Wellington R. Burt, "Forest Lands of the United States Report of the Joint Congressional Committee on Forestry," *Journal of Forestry* 39, no. 4 (1941): 349–52; Lawrence S. Hamilton, "The Federal Forest Regulation Issue: A Recapitulation," *Forest History* 9, no. 1 (1965): 2–11; Rajala, *Clearcutting*; Richardson, *David T. Mason*.

64   Thurman Arnold, quoted in Hidy, Hill, and Nevins, *Timber and Men*, 446. See also Alan Brinkley, "The Antimonopoly Ideal and the Liberal State: The

Case of Thurman Arnold," *Journal of American History* 80, no. 2 (1993): 557–79.

65   For the effects of the dissolution of the NRA on corporate relations with the Roosevelt administration, see Ewen, *P.R.* For the Weyerhaeuser role in the "consent decree," see Hidy, Hill, and Nevins, *Timber and Men.*

66   Zon to Ahern, April 10, 1934, 7 RZP.

67   Zon to Ahern, May 14, 1935, 1/10/7 RMP. This is not to say that he shied from confrontation entirely, as he fought viciously with some of his fellow foresters, including Aldo Leopold, over their trips to Nazi Germany in 1935 and 1936. See, for example, Zon to Mrs. Ahern, July 23, 1935; Zon to Earle H. Clapp, June 7, 1935; Zon to Silcox, May 18, 1936, Zon to Marshall, June 4, 1936, and others, 3–7 RZP. For a description of the American foresters' trips to Germany, see Ward Shepard, "The Purpose of the European Forestry Tour," *Journal of Forestry* 33 (1935): 5–8; Meine, *Aldo Leopold*; and Leopold's Germany files, 3/010/025 ALP.

68   Ahern's health was in decline by the mid-1930s, and he died in 1942. "Col. Ahern, Soldier-Forester, Dies," *American Forests* 48 (June 1942): 276.

69   Zon to Ahern, May 14, 1935, 1/10/7 RMP.

70   For details of Pinchot's post-1934 activities, see Miller, *Gifford Pinchot.*

71   Leopold to Marshall, May 7, 1934, 1/3/14 RMP. While Leopold was an early, high-profile member of The Wilderness Society, he repeatedly declined to play an active role in planning or leadership and wrote of it somewhat ambivalently both before and after Marshall's death; for example, "Defenders of Wilderness," in Leopold, *A Sand County Almanac*, 199–201.

72   Marshall to Lincoln Ellison, April 30, 1935, 1/6/8 RMP, 1.

73   Ibid. Indeed, as one who had worked for years for the Office of Indian Affairs, hearing innumerable stories about prehistoric dwellers on the American landscape, Marshall was more aware than most that by the twentieth century there was no place on the continent untouched by humans. A precise definition of wilderness was not the real business of Marshall's Wilderness Society; protecting that which tended toward the category was. The definition of wilderness, of course, eventually becomes a contentious issue when the definition dictates policy decisions. Critics have often characterized "wilderness" as such an extreme category that it verges on the absurd, but any notion that Marshall himself advocated a completely hands-off wilderness is far from accurate. Nash, *Wilderness and the American Mind*; Cronon, "The Trouble with Wilderness;" Max Oelschlaeger, *The Idea of Wilderness: From Prehistory to the Age of Ecology* (New Haven, CT: Yale University Press, 1991).

74   Marshall to Zon, December 31, 1937, 8 RZP.

75   Mark Harvey, *Wilderness Forever: Howard Zahniser and the Path to the Wilderness Act* (Seattle: University of Washington Press, 2005); Sutter,

Driven Wild; Montgomery, "Evolution of the Society"; Greeley, *Forests and Men*.

76 [Henry Schmitz], "Editorial: The Forest Industries Ask for a Vote of Confidence," *Journal of Forestry* 39 (October 1941): 815–16.

77 Ibid.

Chapter 6

1 For an excellent treatment of the ramifications of sustained-yield forest management plans, see Hirt, *Conspiracy of Optimism*. See also Langston, *Forest Dreams*; Hidy, Hill, and Nevins, *Timber and Men*; and William G. Robbins, *Hard Times in Paradise: Coos Bay, Oregon* (Seattle: University of Washington Press, 2006).

2 Louter, *Windshield Wilderness*; Sutter, *Driven Wild*. See also Olzendam, *Green Gold*; Brock, "Tree Farms on Display"; and Emily K. Brock, "Clemons Tree Farm, Weyerhaeuser Company," *Public Historian* 33, no. 1 (Winter 2011).

3 William W. Grogan, "History of the Establishment of the Clemons Tree Farm," internal history dated February 15, 1943, 7/Public Affairs/Clemons 1-10 WCP.

4 Ibid.; Walker B. Tilley, "American Tree Farms," *Journal of Forestry* 42 (1944): 796–99; Paul F. Sharp, "The Tree Farm Movement," *Agricultural History* 23 (January 1949): 41–45; Hidy, Hill, and Nevins, *Timber and Men*.

5 "Brief Review of the Weyerhaeuser Timber Company and its Individual Operations in Oregon and Washington," USFS Review Paper, October 26, 1938, 3/Correspondence/C. H. Ingram 1938–1939 WCP.

6 Ibid.

7 Winters, *Fifty Years of Forestry*; Hidy, Hill, and Nevins, *Timber and Men*; Clepper, *Professional Forestry*; Robbins, *American Forestry*; Sensel, *Traditions through the Trees*; William Boyd and Scott Prudham, "Manufacturing Green Gold: Industrial Tree Improvement and the Power of Heredity in the Postwar United States," in *Industrializing Organisms: Introducing Evolutionary History*, ed. Susan Schrepfer and Philip Scranton (New York: Routledge, 2004), 107–39. For accounts of the labor disputes between loggers and foresters see, for example, disputes between the Society of American Foresters and various labor unions, 21 SAF.

8 They saw the most promising areas for study at the Clemons to be "rodent control," "improved planting practices," "direct seeding," "fire danger rating," and "development of markets for minor forest products" such as red alder and other brush species. Perry O. Donaldson Jr. and William W. Grogan, "1941 Annual Forestry Report, Clemons Tree Farm," memorandum, May 28, 1942, 7/Public Affairs/Clemons 1-10 WCP.

9 Tilley, "American Tree Farms."

10   "Table 3: Indexes of Effective Fire Control for Various Forest Types," in
     US Department of Agriculture, *National Plan for American Forestry*, 1399;
     Hidy, Hill, and Nevins, *Timber and Men*; Henry Clepper, "Tree Farming in
     America," Unasylva 21 (1967): 3–8; Clepper, *Professional Forestry*.
11   William H. Price, "All-Out Defense of the Clemons Tree Farm," speech at
     the June 12, 1941, Clemons Tree Farm meeting, 4/Branches/Cosmopolis
     WCP; Grogan, "History of the Establishment;" Hidy, Hill, and Nevins,
     Timber and Men.
12   Grogan, "History of the Establishment," 2.
13   Ibid.
14   Sharp, "Tree Farm Movement," 44.
15   Lynn F. Cronemiller, "Oregon's Forest Fire Tragedy: $200,000,000 Loss
     as Flames Destroy Finest Stand of Virgin Timber Remaining in State,"
     American Forests 39 (November 1933): 487–90, 523; William B. Greeley,
     "Forest Fire: The Red Paradox of Conservation," *American Forests* 45
     (April 1939); William G. Morris, "Forest Fires in Western Washington and
     Western Oregon," *Oregon Historical Quarterly* 35 (1934): 313–39; Pyne,
     Fire in America; Wells, *Created Forest*. For the Weyerhaeuser companies'
     views on fire suppression, see Hidy, Hill, and Nevins, *Timber and Men*.
16   The costs broke down as follows: $12,708.58 on initial site improvements:
     roads and trails, buildings and communication systems, and initial
     reforestation; $34,574.32 on initial equipment: tanker and pumper trucks,
     firefighting equipment, and other tools; $9,123.48 for regular maintenance
     costs: fire suppression, debris and brush removal, and continuing reforestation
     expenses. Total costs: $56,406.38. "Comparison of Budget and Accumulated
     Costs to September 30, 1941," 7/Forestry/Clemons Dedication WCP.
17   Roderic Olzendam to J. P. Weyerhaeuser Jr., May 1, 1941, 7/Forestry/
     Clemons Dedication WCP.
18   "Clemons Tree Farm Advertising—1941," manuscript signed GME, March
     17, 1942, 7/Forestry/Clemons Dedication WCP.
19   See, for example, William W. Grogan and W. H. Price, "The Clemons
     Tree Farm: The Detailed and Official Plan of a Vast Private Reforestation
     Project Embracing 130,000 Acres in the Grays Harbor Region," *West
     Coast Lumberman*, July 1941, 12–18, 69–70; "Special Tree Farm Edition,"
     Montesano Vidette, June 12, 1941.
20   Olzendam, "'We the People,'" 1.
21   Ibid., 3.
22   Ibid., 4.
23   Roderic Olzendam, "We Are Growing Trees," *Journal of Forestry* 40, no. 5
     (1942): 393–97.
24   Olzendam, "Timber Is a Crop."
25   Other examples of Olzendam's use of this rhetorical technique can be found
     in Olzendam, "'We the People'"; and Olzendam, "We Are Growing Trees."

26   "Guest Register at Clemons Tree Farm, 1941," 7/Forestry/Clemons Dedication WCP.

27   Untitled informational release to the Montesano Chamber of Commerce for the Grays Harbor County area informational meeting of July 17, 1941, 7/ Forestry/Clemons Dedication WCP. Emphasis in original.

28   "Clemons Tree Farm, Emla, Washington," [pamphlet guide for tourists], Weyerhaeuser Timber Company, n.d. (circa 1951), 32/30 IFA.

29   Brock, "Tree Farms on Display"; Brock, "Clemons Tree Farm"; Chapin Collins, "Industrial Forestry Associations," in *Trees*, USDA Yearbook of Agriculture (Washington, DC: Government Printing Office, 1947), 666–75; W. B. Sayers, "To Tell the Truth: 25 Years of American Forest Products Industries, Inc.," *Journal of Forestry* 64, no. 10 (1966): 657–63; Twining, *Phil Weyerhaeuser*.

30   For Weyerhaeuser Timber Company alone, not including any allied Weyerhaeuser companies, net income in 1941 was $8,613,529.34 and total assets were reported at $153,119,415.87. Weyerhaeuser Timber Company annual report for 1941, published April 1942, 2/Financial/Weyerhaeuser Timber Company annual report WCP.

31   Ibid.

32   Lloyd Thorpe to C. C. Heritage, July 16, 1942, WCR.

33   E. W. Davis memorandum, "Longview Laboratory Publicity," July 23, 1942, WCR.

34   "I understand you are somewhat short of electrical energy during the day, which influenced a decision we made here yesterday to carry on all of the defibrator work in connection with Western woods at Cloquet [Minnesota] instead of Longview until more energy is available, or it appears worthwhile to run in an additional high line to the laboratory." E. W. Davis to Harry Morgan, November 7, 1942, WCR.

35   E. W. Davis to Harry Morgan, November 7, 1942. See also E. W. Davis to J. P. Weyerhaeuser Jr., December 7, 1942, both WCR.

36   The NLMA changed its name to the National Forest Products Association in 1962 and changed it again in 1993 to its current name, the American Forest and Paper Association.

37   "Objectives of the Committee on Public Affairs," NLMA meeting in Chicago, September 11, 1941. Quoted widely, including Wilson Compton to Henry Schmitz, March 7, 1944, as reprinted in "Correspondence," *Journal of Forestry* 42, no. 4 (1944): 301–4. See also Sayers, "To Tell the Truth"; and Collins, "Industrial Forestry Associations."

38   Sharp, "Tree Farm Movement."

39   [Samuel Trask Dana], "Editorial: Why Not a Society Seal of Approval for Acceptable Private Forestry Practices?" *Journal of Forestry* 40, no. 5 (May 1942): 361–62.

40   Ibid., 361.

41   Chronologically, the Clemons Tree Farm was not the first Tree Farm planted.
     A site in Alabama had already been established in a manner that turned out
     to fulfill the criteria set for certification. Because Weyerhaeuser essentially
     controlled publicity on tree farming, this southeastern Tree Farm was not
     generally recognized. Corporate literature regularly referred to the Clemons
     as the nation's first Tree Farm, although Weyerhaeuser publicity knew the
     Alabama site had been first to receive AFPI certification. Indeed, the AFPI,
     through its close relationship with the Northwest lumber industry in general
     and the Weyerhaeuser companies in particular, was largely responsible for
     the ruse. The managing director of the AFPI was Chapin Collins, editor of
     the *Montesano Vidette* and one of the first Tree Farm supporters. Collins
     later wrote that in 1943, "I discovered that actually the first officially
     certified Tree Farm was in Alabama. Since there seemed to be no poetic
     justice in this, we managed to get Clemons referred to in all our publicity as
     [Tree Farm] No. 1. Don't let the south hear about this." Chapin Collins to J.
     P. Weyerhaeuser Jr., 1948. Quoted in Twining, *Phil Weyerhaeuser*, 232.

42   Not only were there more individual Tree Farms certified in Washington
     during these early years, but the acreage put into Tree Farms in Washington
     was more than two and a half times as large as that in Oregon. "Certified
     West Coast Tree Farms: Douglas Fir Region of Washington," 20 NFPA;
     "Certified West Coast Tree Farms: Douglas Fir Region of Oregon," 20
     NFPA.

43   Klamath Falls (OR) *Herald and News*, "Weyerhaeuser Tree Farm Is Nation's
     Top Tree Farm," August 5, 1943.

44   Thomas J. Orr, "Selective Marking in Ponderosa Pine on a Klamath Falls
     Tree Farm," *Journal of Forestry* 43, no. 10 (1945): 738–41.

45   Remarks of the Oregon state forester, Nelson S. Rogers, at the dedication
     ceremony of the Klamath Falls Tree Farm, August 5, 1943, 2 SFC; Chapin
     D. Foster (Pacific Coast Manager of the AFPI) to Nelson S. Rogers, June
     12, 1943, 2 SFC. For a description of the Klamath Falls mill, see Hidy, Hill,
     and Nevins, *Timber and Men*. The classic account of the failures of forest
     improvement in the inland Northwest is in Langston, *Forest Dreams*.

46   Agreement of the AFPI and NLMA, February 28, 1945. Quoted in Sayers,
     "To Tell the Truth."

47   Collins, "Industrial Forestry Associations"; Sayers, "To Tell the Truth";
     Clepper, *Professional Forestry*; Robbins, *American Forestry*.

48   Collins, "Industrial Forestry Associations." The AFPI produced promotional
     booklets such as *Millicoma Forest*, 32/30 IFA; *Men Who Grow Trees*, 32/31
     IFA; and *Cash Crops from Your Woodlands: More Trees for America*, 32/30
     IFA. See Brock, "Tree Farms on Display."

49   Collins, "Industrial Forestry Associations," 2.

50 "Statement of Forest Policy of the National Lumber Manufacturers Association," October 26, 1949. Revised and approved by the board of directors of the NLMA, 20/1 NFPA, 1–2. Capitalization as in original.

51 Ibid., 1.

52 J. P. Weyerhaeuser to Arthur B. Langlie, April 10, 1941, 3/Correspondence/J. P. Weyerhaeuser Jr. WCP, 2.

53 Ibid., 1.

54 Weyerhaeuser Timber Company Annual Report for 1941, published April 1942, 2/Financial /Weyerhaeuser Timber Company annual report WCP.

55 [Samuel Trask Dana], "Editorial: "What's in a Name?'" *Journal of Forestry* 40, no. 8 (1942): 596.

56 Governor Langlie also served on the board, although he was absent from the meeting in question.

57 Minutes of the meeting of the State Forest Board of Washington, June 17, 1941, MMSFB-1941.

58 Ibid., 173.

59 Olzendam, "'We the People,'" 3. Emphasis in original.

60 Minutes of the meeting of the State Forest Board of Washington, June 17, 1941, MMSFB-1941, 160.

61 Ibid., 162.

62 Ibid., 161.

63 W. D. Hagenstein, "Thirty Years of Industrial Tree Farming," transcript of speech given on April 15, 1971, 4/Branches/Cosmopolis WCP.

64 Sharp, "Tree Farm Movement."

65 [Dana], "'What's in a Name?'"

66 Ibid.; Rajala, *Clearcutting*; Clepper, *Professional Forestry*.

67 [Dana], "'What's in a Name?'"

68 Sharp, "Tree Farm Movement"; Brock, "Tree Farms on Display"; Brock, "Clemons Tree Farm."

69 Lyle F. Watts, "Comprehensive Forest Policy Indispensable," *Journal of Forestry* 41 (1943): 783–88; Lyle F. Watts, "A Forest Program to Help Sustain Private Enterprise," *Journal of Forestry* 42 (1944): 81–84; Lyle F. Watts, "The Need for the Conservation of Our Forests," *Journal of Forestry* 42 (1944): 108–14.

70 Lyle F. Watts, *Report of the Chief of the Forest Service* (Washington, DC: Government Printing Office, 1945), 13.

71 Lyle F. Watts, *Annual Report of the Chief of the United States Forest Service* (Washington, DC: Government Printing Office, 1947).

72 C. Edward Behre, "Forest Industry Spreads Dangerous Assumptions on Annual Growth," *Journal of Forestry* 42 (1944): 17–22.

73 Clepper, *Professional Forestry*; Clary, "What Price Sustained Yield?". See also Hirt, *Conspiracy of Optimism*, for later fallout from the shift toward industry.

74   James P. Selvage to Frederick K. Weyerhaeuser, February 13, 1949, 20 WFP,
      1.

Conclusion

1   Leopold, *Sand County Almanac*, 68.
2   Ibid.
3   Ibid., 70.
4   For the nineteenth century and earlier in these forests, see especially White,
      Land Use; and Ficken, *Forested Land*.
5   Hirt, *Conspiracy of Optimism*; Samuel P. Hays, *Wars in the Woods: The Rise
      of Ecological Forestry in America* (Pittsburgh: University of Pittsburgh Press,
      2006). For an examination of this on the Canadian side of the border, see
      Rajala, *Clearcutting*; and Bruce Braun, *The Intemperate Rainforest: Nature,
      Culture, and Power on Canada's West Coast* (Minneapolis: University of
      Minnesota Press, 2002).
6   "Weyerhaeuser Timber Company: History and Nature of Business," 1951,
      127/23 WFP.
7   Ibid.
8   See Brock, "Tree Farms on Display."
9   Marshall to Lincoln Ellison, April 30, 1935, 1/6/8 RMP, 1.
10   Turner, *Promise of Wilderness*; Marsh, *Drawing Lines in the Forest*; Harvey,
      Wilderness Forever.
11   From the preamble to the Wilderness Act of 1964; see Turner, *Promise of
      Wilderness*.
12   The research value of wilderness areas was not a significant part of the
      debate on the final Wilderness Act, although it was widely acknowledged
      that ecologists valued pristine ecosystems as both study areas and baselines
      for comparison with other sites. In many cases federal designation of
      wilderness obstructed scientific research through prohibition and regulation
      of scientists' equipment, access, and experiments.
13   It is worth noting that management practices like thinning and salvage still
      occur in most protected public forestlands in the United States.
14   Kingsland, *Evolution of American Ecology*; see also Worster, *Nature's
      Economy*; and McIntosh, *Background of Ecology*.
15   Rachel Carson, *Silent Spring* (Boston: Houghton Mifflin, 1962); William
      Souder, *On a Farther Shore: The Life and Legacy of Rachel Carson, Author
      of Silent Spring* (New York: Crown Publishing, 2012).
16   Hirt, *Conspiracy of Optimism*; Hays, *Wars in the Woods*.
17   Jerry Franklin's career is described in William Dietrich, *The Final Forest:
      The Battle for the Last Great Trees of the Pacific Northwest* (New York:
      Penguin Books, 1993).

18   The Wind River Experimental Forest under Franklin's direction is described in Herring and Greene, *Forest of Time.*

19   Franklin, quoted in Dietrich, *Final Forest,* 104.

20   Golley, *History of the Ecosystem Concept*; Leslie A. Real and James H. Brown, *Foundations of Ecology: Classic Papers with Commentaries* (Chicago: University of Chicago Press, 1991); Kingsland, *Evolution of American Ecology.*

21   The contentious rise of ecological forestry in the latter half of the twentieth century is well documented in Hays, *Wars in the Woods.* See also Hirt, *Conspiracy of Optimism*; and William G. Robbins, *Landscapes of Conflict: The Oregon Story, 1940–2000* (Seattle: University of Washington Press, 2004).

22   Peter Alagona, *After the Grizzly: Endangered Species and the Politics of Place in California* (Berkeley: University of California Press, 2013); Kurkpatrick Dorsey, *The Dawn of Conservation Diplomacy: U.S.-Canadian Wildlife Protection Treaties in the Progressive Era* (Seattle: University of Washington Press, 1998); Benjamin Kline, *First along the River: A Brief History of the U.S. Environmental Movement,* 4th ed. (Lanham, MD: Rowman and Littlefield, 2011); Robert Gottlieb, *Forcing the Spring: The Transformation of the American Environmental Movement,* rev. ed. (Washington, DC: Island Press, 2005); Turner, *Promise of Wilderness.*

23   An example of intergenerational clashes in policy in another forest is described in Langston, *Forest Dreams.* For discussion of reading landscapes, see Wessels, *Forest Forensics*; Wessels, *Reading the Forested Landscape*; and Cronon, "How to Read a Landscape."

24   Richardson, *BLM's Billion Dollar Checkerboard*; Lawrence E. Davies, "U.S. Policy at Core of Timber Dispute," *New York Times,* April 8, 1951.

25   "The final distribution called for 50 percent of the timber receipts from O&C and Coos Bay Wagon Road lands to go to western Oregon counties, 25 percent to the federal treasury, and 25 percent to the US Forest Service or the BLM for reforestation." *Oregon Blue Book: Almanac and Fact Book 2007–2008* (Corvallis: Oregon State Archives, 2007), 369. See also Paul J. Culhane, *Public Lands Politics: Interest Group Influence on the Forest Service and the Bureau of Land Management* (Baltimore: Resources for the Future Press, 1981); E. Thomas Tuchmann and Chad T. Davis, *O&C Lands Report,* February 6, 2013, http://oregon.gov/gov/GNRO/docs/OCLandsReport.pdf.

26   US Fish and Wildlife Service, "50 CFR Part 17: Endangered and Threatened Wildlife and Plants; Determination of Threatened Status for the Northern Spotted Owl; Final Rule," US Department of the Interior Federal Register 55, no. 123 (June 26, 1990): 26113–26195; Russell Lande, "Demographic Models of the Northern Spotted Owl (*Strix occidentalis caurina*)," *Oecologia*

75 (1988): 601–7. See also Dietrich, *Final Forest*; Robbins, *Landscapes of Conflict*; Barry R. Noon and Kevin S. McKelvey, "Management of the Spotted Owl: A Case History in Conservation Biology," *Annual Review of Ecology and Systematics* 27 (1996): 135–62; Susan Harrison, Andy Stahl, and Daniel Doak, "Spatial Models and Spotted Owls: Exploring Some Biological Issues behind Recent Events," *Conservation Biology* 7, no. 4 (1993). See also US Fish and Wildlife Service, *Recovery Plan for the Northern Spotted Owl* (Portland, OR: US Fish and Wildlife Service, 2008); and US Fish and Wildlife Service, *Revised Recovery Plan for the Northern Spotted Owl* (Portland, OR: US Fish and Wildlife Service, 2011).

27 Noon and McKelvey, "Management of the Spotted Owl"; Tuchmann and Davis, *O&C Lands*.

Bibliography

Ahern, George P. *Deforested America: Statement of the Present Forest Situation in the United States.* Washington, DC: privately printed, 1928.

Ahern, George P., Carlos G. Bates, Earle H. Clapp, L. F. Kneipp, W. C. Lowdermilk, Robert Marshall, E. N. Munns, Gifford Pinchot, Edward C. M. Richards, F. A. Silcox, William N. Sparhawk, and Raphael Zon. "The Petition of June 13, 1934." *Journal of Forestry* 32, no. 7 (1934): 781–83.

Ahern, George P., Robert Marshall, E. N. Munns, Gifford Pinchot, Ward Shepard, W. N. Sparhawk, and Raphael Zon. "A Letter to Foresters." *Journal of Forestry* 28 (April 1930): 456–70.

Alagona, Peter. *After the Grizzly: Endangered Species and the Politics of Place in California.* Berkeley: University of California Press, 2013.

Allen, Garland E. "T. H. Morgan and the Split between Embryology and Genetics, 1910–1935." In *A History of Embryology*, edited by T. J. Horder, J. A. Witkowski, and C. C. Wylie, 113–146. Cambridge: Cambridge University Press, 1985.

Allen, Shirley W. "The Society of American Foresters." In *Fifty Years of Forestry in the U.S.A.*, edited by Robert K. Winters, 272–84. Washington, DC: Society of American Foresters, 1950.

Anker, Peder. *Imperial Ecology: Environmental Order in the British Empire, 1895–1945.* Cambridge, MA: Harvard University Press, 2002.

Barber, William J. *Design within Disorder: Franklin D. Roosevelt, the Economists, and the Shaping of American Economic Policy.* Cambridge: Cambridge University Press, 1996.

Barrow, Mark. *Nature's Ghosts: Confronting Extinction from the Age of Jefferson to the Age of Ecology.* Chicago: University of Chicago Press, 2009.

Bates, Carlos G., and Raphael Zon. *Research Methods in the Study of Forest Environment*, Bulletin of the US Department of Agriculture, no. 1059. Washington, DC: Government Printing Office, 1922.

Behre, C. Edward. "Forest Industry Spreads Dangerous Assumptions on Annual Growth." *Journal of Forestry* 42 (1944): 17–22.

Beidleman, Richard G. *California's Frontier Naturalists.* Berkeley: University of California Press, 2006.

Berman, Gerald S. "Reaction to the Resettlement of World War II Refugees in Alaska." *Jewish Social Studies* 44, no. 3/4 (1982): 271–82.

Best, Gary Dean. *Pride, Prejudice, and Politics: Roosevelt versus Recovery.* New York: Praeger, 1991.

Bevir, Mark. *The Making of British Socialism.* Princeton, NJ: Princeton University Press, 2011.

Bocking, Stephen. *Nature's Experts: Science, Politics, and the Environment.* New Brunswick, NJ: Rutgers University Press, 2004.

Bont, Raf de. "Evolutionary Morphology in Belgium: The Fortunes of the 'Van Beneden School,' 1870–1900." *Journal of the History of Biology* 41, no. 1 (2008): 81–118.

Bowers, Janice E. *A Sense of Place: The Life and Work of Forrest Shreve.* Tucson: University of Arizona Press, 1988.

Boyd, William, and Scott Prudham. "Manufacturing Green Gold: Industrial Tree Improvement and the Power of Heredity in the Postwar United States." In *Industrializing Organisms: Introducing Evolutionary History,* edited by Susan Schrepfer and Philip Scranton, 107–39. New York: Routledge, 2004.

Bradley, Richard. *General Treatise of Husbandry and Gardening, Containing a New System of Vegetation.* London: T. Woodward, 1726.

Brant, Irving. *Adventures in Conservation with Franklin D. Roosevelt.* Flagstaff, AZ: Northland Publishing, 1988.

Braun, Bruce. *The Intemperate Rainforest: Nature, Culture, and Power on Canada's West Coast.* Minneapolis: University of Minnesota Press, 2002.

Brinkley, Alan. "The Antimonopoly Ideal and the Liberal State: The Case of Thurman Arnold." *Journal of American History* 80, no. 2 (1993): 557–79.

Brinkley, Douglas. *The Quiet World: Saving Alaska's Wilderness Kingdom, 1879–1960.* New York: HarperCollins, 2011.

Brock, Emily K. "The Challenge of Reforestation: Ecological Experiments in the Douglas Fir Forest, 1920–1940." *Environmental History* 9, no. 1 (2004): 57–79.

———. "Clemons Tree Farm, Weyerhaeuser Company." *Public Historian* 33, no. 1 (Winter 2011).

———. "New Patterns in Old Places: Forest History for the Global Present." In *The Oxford Handbook of Environmental History,* edited by Andrew C. Isenberg, 154–77. New York: Oxford University Press, 2014.

———. "Tree Farms on Display: Presenting Industrial Forests to the Public in the Pacific Northwest, 1941–1960." *Oregon Historical Quarterly* 113 (Winter 2012): 526–59.

Browne, Janet. *Darwin's Origin of Species: A Biography.* New York: Atlantic Monthly Press, 2006.

Burt, Wellington R. "Forest Lands of the United States Report of the Joint Congressional Committee on Forestry." *Journal of Forestry* 39, no. 4 (1941): 349–52.

Büsgen, M. *The Structure and Life of Forest Trees.* 3rd ed. Revised by E. Münch. Translated by Thomas Thomson. London: Chapman and Hall, 1929.

Carson, Rachel. *Silent Spring.* Boston: Houghton Mifflin, 1962.

[Chamberlin, Thomas Chrowder]. "Editorial." *Journal of Geology* 18, no. 5 (1910): 468–69.

Chapman, H. H. "Making the Society of American Foresters a Professional Organization." *Journal of Forestry* 32, no. 4 (1934): 503–7.

———. "The Responsibilities of the Profession in the Present Situation." *Journal of Forestry* 33, no. 3 (1935): 204–14.

———. "Second Conference on the Lumber Code." *Journal of Forestry* 32, no. 3 (1934): 272–307.

———. "Statement by President Chapman." *Journal of Forestry* 32, no. 7 (October 1934): 777–81.

[Chapman, H. H.]. "Editorial: Professional Idealism." *Journal of Forestry* 32, no. 7 (1934): 677–79.

Churchill, Frederick B. "Weismann, Hydromedusae, and the Biogenetic Imperative: A Reconsideration." In *A History of Embryology*, edited by T. J. Horder, J. A. Witkowski, and C. C. Wylie, 7–33. Cambridge: Cambridge University Press, 1985.

Cittadino, Eugene. *Nature as the Laboratory: Darwinian Plant Ecology in the German Empire, 1880–1900.* New York: Cambridge University Press, 1990.

Clary, David A. "What Price Sustained Yield? The Forest Service, Community Stability, and Timber Monopoly under the 1944 Sustained-Yield Act." *Journal of Forest History* 31, no. 1 (January 1987): 4–18.

Clements, Edith S. *Adventures in Ecology.* New York: Pageant Press, 1960.

Clements, Frederic E. *Life History of Lodgepole Burn Forests*, USDA Forest Service Bulletin 79. Washington, DC: Government Printing Office, 1910.

———. *Research Methods in Ecology.* Lincoln, NE: University Publishing Company, 1905.

Clepper, Henry. *Professional Forestry in the United States.* Baltimore: Resources for the Future Press / The Johns Hopkins University Press, 1971.

———. "Tree Farming in America." *Unasylva* 21 (1967): 3–8.

"Col. Ahern, Soldier-Forester, Dies." *American Forests* 48 (June 1942): 276.

Cole, G. D. II. *British Working Class Politics, 1832–1914.* London: G. Routledge and Sons, 1941.

Colinvaux, Paul A. "The Efficiency of Life." In *Why Big Fierce Animals Are Rare: An Ecologist's Perspective*, 32–46. Princeton, NJ: Princeton University Press, 1978.

Collins, Chapin. "Industrial Forestry Associations." In *Trees*, USDA Yearbook of Agriculture, 666–75. Washington, DC: Government Printing Office, 1947.

Cornell, Thomas D. "Carnegie Institution of Washington." In *History of Science in the United States: An Encyclopedia*, edited by Marc Rothenberg, 102–8. New York: Garland, 2001.

Coville, Frederick Vernon, and Daniel Trembly MacDougal. *Desert Botanical Laboratory of the Carnegie Institution.* Washington, DC: Carnegie Institution, 1903.

Cowles, Henry Chandler. "An Ecological Aspect of the Conception of Species." *American Naturalist* 42, no. 496 (1908): 265–71.

———. "The Ecological Relations of the Vegetation on the Sand Dunes of Lake Michigan. Part I: Geographical Relations of the Dune Floras." *Botanical Gazette* 27, no. 2 (1899): 95–117.

Craig, Patricia. *Centennial History of the Carnegie Institution of Washington.* Vol. 4, The Department of Plant Biology. New York: Cambridge University Press, 2005.

Craige, Betty Jean. *Eugene Odum: Ecosystem Ecologist and Environmentalist.* Athens: University of Georgia Press, 2001.

Cronemiller, Lynn F. "Oregon's Forest Fire Tragedy: $200,000,000 Loss as Flames Destroy Finest Stand of Virgin Timber Remaining in State." *American Forests* 39 (November 1933): 487–90, 523.

Cronon, William. "Kennecott Journey: The Paths Out of Town." In *Under an Open Sky: Rethinking America's Western Past*, edited by William Cronon, George Miles, and Jay Gitlin, 28–51. New York: W. W. Norton, 1992.

———. *Nature's Metropolis: Chicago and the Great West.* New York: W. W. Norton, 1992.

———. "The Trouble with Wilderness; or, Getting Back to the Wrong Nature." In *Uncommon Ground: Rethinking the Human Place in Nature*, edited by William Cronon, 69–90. New York: W. W. Norton, 1995.

Culhane, Paul J. *Public Lands Politics: Interest Group Influence on the Forest Service and the Bureau of Land Management.* Baltimore: Resources for the Future Press, 1981.

Culver, John C., and John Hyde, *American Dreamer: The Life and Times of Henry A. Wallace.* New York: W. W. Norton, 2000.

[Dana, Samuel Trask]. "Editorial: 'What's in a Name?'" *Journal of Forestry* 40, no. 8 (1942): 596.

[Dana, Samuel Trask]. "Editorial: Why Not a Society Seal of Approval for Acceptable Private Forestry Practices?" *Journal of Forestry* 40, no. 5 (May 1942): 361–62.

Darwin, Charles. *On the Origin of Species by Means of Natural Selection, or the Preservation of Favoured Races in the Struggle for Life.* 1865. Facsimile of the first edition, with an introduction by Ernst Mayr. Cambridge, MA: Harvard University Press, 1964.

Darwin, Francis. *The Life and Letters of Charles Darwin.* New York: Appleton, 1898.

Dickinson, Matthew J. *Bitter Harvest: FDR, Presidential Power, and the Growth of the Presidential Branch.* Cambridge: Cambridge University Press, 1997.

Dietrich, William. *The Final Forest: The Battle for the Last Great Trees of the Pacific Northwest.* New York: Penguin Books, 1993.

Doig, Ivan. *Early Forestry Research: A History of the Pacific Northwest Forest and Range Experiment Station, 1925–1975.* Washington, DC: USDA Forest Service, 1976.

Dorsey, Kurkpatrick. *The Dawn of Conservation Diplomacy: U.S.-Canadian Wildlife Protection Treaties in the Progressive Era.* Seattle: University of Washington Press, 1998.

Ebert, James D. "Carnegie Institution of Washington and Marine Biology: Naples, Woods Hole, and Tortugas." *Biological Bulletin* 168 (1985): 172–82.

Elton, Charles S. *Animal Ecology.* London: Sidgwick and Jackson, 1927.

Ewen, Stuart. *P.R.: A Social History of Spin.* New York: Basic Books, 1996.

Farmer, Jared. *Trees in California: A California History.* New York: W. W. Norton, 2013.

Ficken, Robert E. *The Forested Land: A History of Lumbering in Western Washington.* Seattle: University of Washington Press, 1987.

Flader, Susan L. *Thinking like a Mountain: Aldo Leopold and the Evolution of an Ecological Attitude toward Deer, Wolves, and Forests.* Columbia: University of Missouri Press, 1974.

Forester, A. [pseud.]. "A Research Forester Looks at His Job." *Journal of Forestry* 32, no. 5 (1934): 598–603.

Fritz, Emanuel. "Report of the Editor: Resignation Tendered." *Journal of Forestry* 31, no. 2 (1933): 233–35.

———. "Statement by Former Editor-in-Chief Emanuel Fritz." *Journal of Forestry* 32, no. 7 (1934): 785.

Ganong, William F. "The Cardinal Principles of Ecology." *Science* 19, no. 482 (March 25, 1904): 493–98.

Gennett, Andrew. *Sound Wormy: Memoir of Andrew Gennett, Lumberman.* Athens: University of Georgia Press, 2002.

Gleason, Henry A. "The Vegetational History of the Middle West." *Annals of the Association of American Geographers* 12 (1922): 39–85.

Glover, James M. A *Wilderness Original: The Life of Bob Marshall.* Seattle: Mountaineers, 1986.

Golley, Frank. *A History of the Ecosystem Concept in Ecology: More Than the Sum of Its Parts.* New Haven, CT: Yale University Press, 1993.

Gottlieb, Robert. *Forcing the Spring: The Transformation of the American Environmental Movement.* 1993. Rev. ed. Washington, DC: Island Press, 2005.

Greeley, William B. "Economic Aspects of Our Timber Supply." *Journal of Forestry* 20, no. 8 (1922): 837–47.

———. "Forest Fire: The Red Paradox of Conservation." *American Forests* 45 (April 1939): 152–57.

———. *Forests and Men.* New York: Doubleday, 1951.

———. "Industrial Forestry." In *Fifty Years of Forestry in the U.S.A.*, edited by Robert K. Winters, 238–59. Washington, DC: Society of American Foresters, 1950.

Greeley, William B., Earle H. Clapp, Joseph Kittredge, Ward Shepard, Herbert A. Smith, William N. Sparhawk, and Raphael Zon. "Timber: Mine or Crop?" In *US Department of Agriculture Yearbook 1922*, 83–180. Washington, DC: Government Printing Office, 1923.

Grogan, William W., and W. H. Price. "The Clemons Tree Farm: The Detailed and Official Plan of a Vast Private Reforestation Project Embracing 130,000 Acres in the Grays Harbor Region." *West Coast Lumberman*, July 1941, 12–18, 69–70.

Haeckel, Ernst. *Natürliche Schöpfungsgeschichte: Gemeinverständliche Wissenschaftliche . . .* [The history of creation, or, the development of earth and its inhabitants . . .]. Translated by Ray Lankester. 1876. Reprint, Cambridge: Chadwyck-Healey, 1990.

Hagen, Joel B. *An Entangled Bank: The Origin of Ecosystem Ecology.* New Brunswick, NJ: Rutgers University Press, 1992.

Hamilton, Lawrence S. "The Federal Forest Regulation Issue: A Recapitulation." *Forest History* 9, no. 1 (1965): 2–11.

Harrison, Susan, Andy Stahl, and Daniel Doak. "Spatial Models and Spotted Owls: Exploring Some Biological Issues Behind Recent Events." *Conservation Biology* 7, no. 4 (1993).

Harvey, Mark. *Wilderness Forever: Howard Zahniser and the Path to the Wilderness Act.* Seattle: University of Washington Press, 2005.

Hays, Samuel P. *The American People and the National Forests.* Pittsburgh: University of Pittsburgh Press, 2009.

———. *Conservation and the Gospel of Efficiency: The Progressive Conservation Movement, 1890–1920.* 1959. Reprint, Pittsburgh: University of Pittsburgh Press, 1999.

———. *Wars in the Woods: The Rise of Ecological Forestry in America.* Pittsburgh: University of Pittsburgh Press, 2006.

Hazard, J. O. "Hap." "The Roosevelt Forestry Song." *American Forests* 39 (October 1933): 250.

Healey, Judith Koll. *Frederick Weyerhaeuser and the American West.* St. Paul: Minnesota Historical Society Press, 2013.

Herring, Margaret, and Sarah Greene. *Forest of Time: A Century of Science at Wind River Experimental Forest.* Corvallis: Oregon State University Press, 2007.

Herron, John P. *Science and the Social Good: Nature, Culture, and Community, 1865–1965.* New York: Oxford University Press, 2010.

Hidy, Ralph W. *The Great Northern Railway: A History.* Boston: Harvard Business School Press, 1988.

Hidy, Ralph W., Frank Ernest Hill, and Alan Nevins. *Timber and Men: The Weyerhaeuser Story.* New York: Macmillan, 1963.

Highsmith, Richard M., Jr., and John L. Beh. "Tillamook Burn: The Regeneration of a Forest." *Scientific Monthly* 75, no. 3 (September 1952): 139–48.

Hirt, Paul. *A Conspiracy of Optimism: Management of the National Forests since World War Two.* Lincoln: University of Nebraska Press, 1994.

Hofmann, J. V. "The Establishment of a Douglas Fir Forest." *Ecology* 1 (1920): 49–53.

———. "Furred Forest Planters." *Scientific Monthly* 16 (1923): 280–83.

———. "Natural Reproduction from Seed Stored in the Forest Floor." *Journal of Agricultural Research* 11 (October 1917): 1–26.

Hogue, Arthur R. "The Carl Schurz Memorial Foundation: The First Twenty-Five Years." *Indiana Magazine of History* 51, no. 4 (1955): 335–39.

Holbrook, Stewart. *Burning an Empire: The Study of American Forest Fires.* New York: Macmillan, 1943.

Hosmer, Ralph A. "Education in Professional Forestry." In *Fifty Years of Forestry in the U.S.A.,* edited by Robert K. Winters, 299–315. Washington, DC: Society of American Foresters, 1950.

Huntington, Ellsworth. "The Secret of the Big Trees." *Harper's Magazine* 125 (July 1912): 292–302.

Ice, George G., and John D. Stednick. "Forest Watershed Research in the United States." *Forest History Today,* Spring/Fall 2004, 16–26.

Igler, David. *The Great Ocean: Pacific Worlds from Captain Cook to the Gold Rush.* New York: Oxford University Press, 2013.

Isaac, Leo A. "Life of Douglas Fir Seed in the Forest Floor." *Journal of Forestry* 33 (1935): 61–66.

———. "Seed Flight in the Douglas Fir Region." *Journal of Forestry* 28 (1930): 492–99.

Isaac, Leo A., and Amelia R. Fry, *Douglas Fir Research in the Pacific Northwest, 1920–1956.* Berkeley: Regional Oral History Office, Bancroft Library, University of California, 1967.

———. "The Seed-Flight Experiment: Policy Heeds Research." *Forest History* 16 (October 1973): 54–60.

Isenberg, Andrew C. *The Destruction of the Bison: An Environmental History, 1750–1920.* New York: Cambridge University Press, 2000.

———. *Mining California: An Ecological History.* New York: Hill and Wang, 2005.

Jensen, Derrick, and George Draffan. *Railroads and Clearcuts: Legacy of Congress's 1864 Northern Pacific Railroad Land Grant.* Spokane, WA: Inland Empire Public Lands Council, 1995.

Juncker, Clara. "Bureaucratic Dynamics and Control of the New Deal's Publicity: Struggles between Core and Periphery in the FSA's Information Division."

In *The Roosevelt Years: New Perspectives on American History, 1933–1945*, edited by Robert Garson and Stuart S. Kidd. Edinburgh: Edinburgh University Press, 1999.

Kellogg, R. S. "As I See It." *Journal of Forestry* 28 (April 1930): 461–62.

Kingsland, Sharon E. *The Evolution of American Ecology, 1890–2000*. Baltimore: The Johns Hopkins University Press, 2005.

———. *Modeling Nature: Episodes in the History of Population Ecology*. Chicago: University of Chicago Press, 1995.

Kjeldsen, Jim. *The Mountaineers: A History*. Seattle: Mountaineers Books, 1998.

Kline, Benjamin. *First Along the River: A Brief History of the U.S. Environmental Movement*. 4th ed. Lanham, MD: Rowman and Littlefield, 2011.

Kneipp, L. F. "Industrial Forestry." *Journal of Forestry* 26, no. 4 (1928): 507–12.

Kohler, Robert E. *Landscapes and Labscapes: Exploring the Lab-Field Border in Ecology*. Chicago: University of Chicago Press, 2002.

———. *Partners in Science: Foundations and Natural Scientists, 1900–1945*. Chicago: University of Chicago Press, 1991.

Lande, Russell. "Demographic Models of the Northern Spotted Owl (*Strix occidentalis caurina*)." *Oecologia* 75 (1988): 601–7.

Langston, Nancy. *Forest Dreams, Forest Nightmares: The Paradox of Old Growth in the Inland West*. Seattle: University of Washington Press, 1995.

Lears, T. J. Jackson. *Fables of Abundance: A Cultural History of Advertising in America*. New York: Basic Books, 1994.

Leopold, Aldo. "Conservation Economics." *Journal of Forestry* 32, no. 5 (1934): 537–44.

———. "Deer and Dauerwald in Germany: I. History." *Journal of Forestry* 34, no. 4 (April 1936): 366–75.

———. *A Sand County Almanac and Sketches Here and There*. New York: Oxford University Press, 1949.

Lewis, David E. "The University of Kazan: Provincial Cradle of Russian Organic Chemistry." *Journal of Chemical Education* 71, no. 1 (1994): 93–97.

Livingston, Burton E. "Research Methods in Ecology." *Ecology* 5, no. 1 (January 1924): 99–101.

"A Lost Leader." *American Forests* 16, no. 10 (1910): 604–6.

Louter, David. *Windshield Wilderness: Cars, Roads, and Nature in Washington's National Parks*. Seattle: University of Washington Press, 2006.

Lucia, Ellis. *Tillamook Burn Country*. Caldwell, ID: Caxton Printers, 1983.

Maher, Neil M. *Nature's New Deal: The Civilian Conservation Corps and the Roots of the American Environmental Movement*. New York: Oxford University Press, 2009.

Malone, Michael P. *James J. Hill: Empire Builder*. Norman: University of Oklahoma Press, 1996.

Markowitz, Norman D. *The Rise and Fall of the People's Century: Henry A. Wallace and American Liberalism, 1941–1948.* New York: Free Press, 1973.

Marsh, Kevin. *Drawing Lines in the Forest: Creating Wilderness Areas in the Pacific Northwest.* Seattle: University of Washington Press, 2010.

Marshall, Robert. "Adventure, Arrogance, and the Arctic." *Wings,* May 1933, 9–12.

———. *Alaskan Wilderness: Exploring the Central Brooks Range.* Edited by George Marshall. Berkeley: University of California Press, 2005.

———. *Arctic Village.* New York: Literary Guild, 1933.

———. *Arctic Wilderness.* Edited and with an introduction by George Marshall. Berkeley: University of California Press, 1956.

———. *Bob Marshall in the Adirondacks: Writings of a Pioneering Peak-Bagger, Pond-Hopper, and Wilderness Preservationist.* Edited by Phil Brown. Saranac Lake, NY: Lost Pond Press, 2006.

———. "Comments on Commission's Truck Trail Policy." *American Forests* 43 (January 1936): 6, 48.

———. "Contribution to the Life History of the Northwestern Lumberjack." *Social Forces* 8 (December 1929): 271–73.

———. "An Experimental Study of the Water Relations of Seedling Conifers with Special Reference to Wilting." PhD diss., Johns Hopkins University, 1930.

———. "Fallacies in Osborne's Position." *Living Wilderness* 1 (1935): 4.

———. "Influence on Precipitation Cycles in Forestry." *Journal of Forestry* 25, no. 4 (1927): 415–29.

———. *The People's Forests.* 1933. Reprint, Iowa City: University of Iowa Press, 2002.

———. "Precipitation and Presidents." *The Nation* 124 (March 23, 1927): 316–17.

———. "The Problem of the Wilderness." *Scientific Monthly* 30, no. 2 (February 1930): 141–48.

———. "Recreational Limitations of Silviculture in the Adirondacks." *Journal of Forestry* 23, no. 2 (1925): 173–78.

———. *The Social Management of American Forests.* New York: League for Industrial Democracy, 1930.

Mason, David T. *Forests for the Future: The Story of Sustained Yield as Told in the Diaries and Papers of David T. Mason, 1907–1950.* Edited by Rodney C. Loeher. St. Paul: Forest Products History Foundation / Minnesota Historical Society, 1950.

Matthew, Patrick. *On Naval Timber and Arboriculture.* 1831. Electronic facsimile edition, http://darwin-online.org.uk/content/frameset?viewtype=side&itemID=A154&pageseq=1.

Maunder, Elwood R. *A Forester's Log: Fifty Years in the Pacific Northwest; An Interview with George L. Drake.* Durham, NC: Forest History Society, 2004.

McIntosh, Robert P. *The Background of Ecology: Concept and Theory.* Cambridge: Cambridge University Press, 1991.

———. "Pioneer Support for Ecology." *BioScience* 33 (February 1983): 107–12.

Meine, Curt. *Aldo Leopold: His Life and Work.* Madison: University of Wisconsin Press, 1988.

Menard, Andrew. *Sight Unseen: How Frémont's First Expedition Changed the American Landscape.* Lincoln: University of Nebraska Press, 2012.

Merrill, Karen R. *Public Lands and Political Meaning: Ranchers, the Government, and the Property between Them.* Berkeley: University of California Press, 2002.

Metz, C. B., ed. "The Naples Zoological Station and the Marine Biological Laboratory: One Hundred Years of Biology," supplement, *Biological Bulletin* 168 (1985): 1–207.

Miller, Char. *Gifford Pinchot and the Making of Modern Environmentalism.* Washington, DC: Island Press, 2001.

———. "Militant Forester, Raphael Zon." *Forest Magazine,* Winter 2005, 22–23.

Mitman, Gregg. *The State of Nature: Ecology, Community, and American Social Thought, 1900–1950.* Chicago: University of Chicago Press, 1992.

Montgomery, David. "Evolution of the Society of American Foresters 1934–1937 as Seen in the Memoirs of H. H. Chapman." *Forest History Newsletter* 6, no. 3 (1962): 2–9.

Morris, William G. "Forest Fires in Western Washington and Western Oregon." *Oregon Historical Quarterly* 35 (1934): 313–39.

Muir, John. "The Hetch Hetchy Valley." *Sierra Club Bulletin* 6, no. 4 (1908): 211–20.

———. "Our National Forests." *Atlantic Monthly* 80, no. 478 (August 1897): 145–57.

Munger, Thornton T., and Amelia R. Fry. *Forest Research in the Northwest.* Berkeley: Regional Oral History Office, Bancroft Library, University of California, 1967.

Nash, Roderick. "The Strenuous Life of Bob Marshall." *Forest History Newsletter* 10, no. 3 (1966): 18–25.

———. *Wilderness and the American Mind.* New Haven, CT: Yale University Press, 1967.

Nichols, Tad, and Bill Broyles. "Afield with Desert Scientists." *Journal of the Southwest* 39 (Autumn-Winter 1997): 353–70.

Noon, Barry R., and Kevin S. McKelvey. "Management of the Spotted Owl: A Case History in Conservation Biology." *Annual Review of Ecology and Systematics* 27 (1996): 135–62.

Odum, Eugene P. *Fundamentals of Ecology.* Philadelphia: W. B. Saunders, 1953.

Oelschlaeger, Max. *The Idea of Wilderness: From Prehistory to the Age of Ecology.* New Haven, CT: Yale University Press, 1991.

Olzendam, Roderic. *Green Gold for America: The Life and Times of Roderic Marble Olzendam*. Portland, OR: Binford and Mort, 1982.

———. "We Are Growing Trees." *Journal of Forestry* 40, no. 5 (1942): 393–97.

Oregon Blue Book: Almanac and Fact Book 2007–2008. Corvallis: Oregon State Archives, 2007.

Orr, Thomas J. "Selective Marking in Ponderosa Pine on a Klamath Falls Tree Farm." *Journal of Forestry* 43, no. 10 (1945): 738–41.

Owen, A. L. Reisch. *Conservation under FDR*. New York: Praeger, 1983.

Pauly, Philip J. *Biologists and the Promise of American Life: From Meriwether Lewis to Alfred Kinsey*. Princeton, NJ: Princeton University Press, 2000.

Pease, Edward R. *History of the Fabian Society*. New York: E. P. Dutton, 1916.

Pinchot, Gifford. *Breaking New Ground*. 1947. Reprint, New York: Island Press, 1998.

———. *A Primer of Forestry, Part I: The Forest*. USDA Division of Forestry Bulletin, no. 24, vol. 1, 1899.

———. *A Primer of Forestry, Part II: Practical Forestry*. USDA Division of Forestry Bulletin, no. 24, vol. 2, 1905.

———. *To the South Seas: The Cruise of the Schooner Mary Pinchot to the Galapagos, the Marquesas, and the Tuamotu Islands, and Tahiti*. Philadelphia: John C. Winston, 1930.

"Proceedings, 34th Annual Meeting of the Society of American Foresters." *Journal of Forestry* 33, no. 3 (1935): 189–90.

Prudham, W. Scott. *Knock on Wood: Nature as Commodity in Douglas-Fir Country*. New York: Routledge, 2005.

Pyne, Stephen. *Fire in America: A Cultural History of Wildland and Rural Fire*. Seattle: University of Washington Press, 1997.

Rajala, Richard. *Clearcutting the Pacific Rain Forest: Production, Science, and Regulation*. Vancouver: University of British Columbia Press, 1998.

Real, Leslie A., and James H. Brown. *Foundations of Ecology: Classic Papers with Commentaries*. Chicago: University of Chicago Press, 1991.

Recknagel, A. B. "Summary of Forest Practice Rules under the Conservation Code (Article X)." *Journal of Forestry* 33, no. 9 (1935): 792–95.

Reed, Franklin. "An Editorial Policy for the *Journal of Forestry*." *Journal of Forestry* 32, no. 7 (1934): 785–91.

Richards, Robert J. *The Tragic Sense of Life: Ernst Haeckel and the Struggle over Evolutionary Thought*. Chicago: University of Chicago Press, 2008.

Richardson, Elmo. *BLM's Billion Dollar Checkerboard: Managing the O&C Lands*. Santa Cruz, CA: Forest History Society, 1980.

———. *David T. Mason, Forestry Advocate: His Role in the Application of Sustained Yield Management to Private and Public Forest Lands*. Durham, NC: Forest History Society, 1983.

Righter, Robert. "National Monuments to National Parks: The Use of the Antiquities Act of 1906." *Western Historical Quarterly* 20 (August 1989): 281–301.

Robbins, William G. *American Forestry: A History of National, State, and Private Cooperation.* Lincoln: University of Nebraska Press, 1985.

———. *Hard Times in Paradise: Coos Bay, Oregon.* Seattle: University of Washington Press, 2006.

———. *Landscapes of Conflict: The Oregon Story, 1940–2000.* Seattle: University of Washington Press, 2004.

———. *Landscapes of Promise: The Oregon Story, 1800–1940.* Seattle: University of Washington Press, 1997.

———. *Lumberjacks and Legislators: Political Economy of the U.S. Lumber Industry, 1890–1941.* College Station: Texas A&M University Press, 1982.

Rodgers, Andrew Denny, II. *Bernhard Eduard Fernow: A Story of North American Forestry.* Princeton, NJ: Princeton University Press, 1951.

Roosevelt, Franklin D. *Franklin D. Roosevelt and Conservation: 1911–1945.* Compiled and edited by Edgar B. Nixon. Hyde Park, NY: Franklin D. Roosevelt Library, 1957.

Roosevelt, Theodore. *African Game Trails: An Account of the African Wanderings of an American Hunter-Naturalist.* New York: Charles Scribner's Sons, 1910.

Rudolph, Paul O. "R. Zon, Pioneer in Forest Research." *Science* 125, no. 3261 (1957): 1283–84.

Salmond, John A. *The Civilian Conservation Corps, 1933–1942: A New Deal Case Study.* Durham, NC: Duke University Press, 1967.

Sartz, Richard S. "Carlos G. Bates: Maverick Forest Service Scientist." *Journal of Forest History* 21, no. 1 (1977): 31–39.

Sayers, W. B. "To Tell the Truth: 25 Years of American Forest Products Industries, Inc." *Journal of Forestry* 64, no. 10 (1966): 657–63.

Schenck, Carl A. *Cradle of Forestry in America: The Biltmore Forest School, 1898–1913.* Edited by Ovid Butler. Durham, NC: Forest History Society, 1998.

Schlesinger, Arthur M., Jr. *The Age of Roosevelt: The Coming of the New Deal.* Boston: Houghton Mifflin, 1958.

Schmaltz, Norman J. "Forest Researcher: Raphael Zon, Part I." *Journal of Forest History* 24, no. 1 (1980): 25–39.

———. "Forest Researcher: Raphael Zon, Part II." *Journal of Forest History* 24, no. 2 (1980): 86–97.

[Schmitz, Henry]. "Editorial: The Forest Industries Ask for a Vote of Confidence." *Journal of Forestry* 39 (October 1941): 815–16.

Schrepfer, Susan R. *The Fight to Save the Redwoods: A History of Environmental Reform, 1917–1978.* Madison: University of Wisconsin Press, 1983.

Sensel, Joni. *Traditions Through the Trees: Weyerhaeuser's First 100 Years.* Seattle: Documentary Book Publishers, 1999.

Sharp, Paul F. "The Tree Farm Movement." *Agricultural History* 23 (January 1949): 41–45.

Shaw, George Bernard. *The Fabian Society: Its Early History.* Fabian Tract no. 41. London: Fabian Society, 1892.

Shepard, Ward. "Cooperative Control: A Proposed Solution of the Forest Problem." *Journal of Forestry* 28 (February 1930): 113–20.

———. "The Necessity for Realism in Forestry Propaganda." *Journal of Forestry* 25 (January 1927): 11–26.

———. "The Purpose of the European Forestry Tour." *Journal of Forestry* 33 (1935): 5–8.

Shepard, Ward, and Franz Heske. "European Facts for American Skeptics." *Journal of Forestry* 31, no. 8 (1933): 923–31.

Silver, M. M. *Louis Marshall and the Rise of Jewish Ethnicity in America: A Biography.* Syracuse: Syracuse University Press, 2013.

Skeel, David A. *Debt's Dominion: A History of Bankruptcy Law in America.* Princeton, NJ: Princeton University Press, 2001.

[Smith, Herbert A.]. "Editorial: The Cult of the Wilderness." *Journal of Forestry* 33 (December 1935): 955–57.

"Society Affairs." *Journal of Forestry* 32, no. 7 (1934): 777–96.

Souder, William. *On a Farther Shore: The Life and Legacy of Rachel Carson, Author of Silent Spring.* New York: Crown Publishing, 2012.

Steen, Harold K. *Forest Service Research: Finding Answers to Conservation's Questions.* Durham, NC: Forest History Society, 1998.

Sutter, Paul S. *Driven Wild: How the Fight against Automobiles Launched the Modern Wilderness Movement.* Seattle: University of Washington Press, 2002.

Thompson, J. Lee. *Theodore Roosevelt Abroad: Nature, Empire, and the Journey of an American President.* New York: Palgrave Macmillan, 2010.

Tilley, Walker B. "American Tree Farms." *Journal of Forestry* 42 (1944): 796–99.

Tobey, Ronald C. *Saving the Prairies: The Life Cycle of the Founding School of American Plant Ecology, 1895–1955.* Berkeley: University of California Press, 1981.

Transeau, Edgar Nelson. "The Accumulation of Energy by Plants." *Ohio Journal of Science* 26, no. 1 (1926): 1–10.

Tuchmann, E. Thomas, and Chad T. Davis. *O&C Lands Report.* Prepared for Oregon Governor John Kitzhaber, February 6, 2013. http://oregon.gov/gov/GNRO/docs/OCLandsReport.pdf.

Tucker, Richard P. "Unsustainable Yields: American Foresters and Tropical Timber Resources." In *Insatiable Appetite: The United States and the Ecological Degradation of the Tropical World*, 345–415. Berkeley: University of California Press, 2000.

Turner, James Morton. *The Promise of Wilderness: American Environmental Politics since 1964.* Seattle: University of Washington Press, 2012.

Twight, Ben W. *Organizational Values and Political Power: The Forest Service and the Olympic National Park*. University Park: Pennsylvania State University Press, 1983.

Twining, Charles E. *A Bundle of Sticks: The Story of a Family*. St. Paul: Rock Island, 1987.

———. *F. K. Weyerhaeuser: A Biography*. St. Paul: Minnesota Historical Society Press, 1997.

———. *George S. Long: Timber Statesman*. Seattle: University of Washington Press, 1994.

———. *Phil Weyerhaeuser, Lumberman*. Seattle: University of Washington Press, 1985.

———. *The Tie That Binds: A History of the Weyerhaeuser Family Office*. St. Paul: Rock Island, 1993.

Tye, Larry. *The Father of Spin: Edward L. Bernays and the Birth of Public Relations*. New York: Crown Publishing, 1998.

US Department of Agriculture. *A National Plan for American Forestry* [Copeland Report]. S. Doc. No. 12, 73rd Cong., 1st sess., March 13, 1933.

US Fish and Wildlife Service. "50 CFR Part 17: Endangered and Threatened Wildlife and Plants; Determination of Threatened Status for the Northern Spotted Owl; Final Rule." *US Department of the Interior Federal Register* 55, no. 123 (June 26, 1990): 26113–26195.

———. *Recovery Plan for the Northern Spotted Owl* (Portland, OR: US Fish and Wildlife Service, 2008).

———. *Revised Recovery Plan for the Northern Spotted Owl* (Portland, OR: US Fish and Wildlife Service, 2011).

Vetter, Jeremy. "Rocky Mountain High Science: Teaching, Research and Nature at Field Stations." In *Knowing Global Environments: New Historical Perspectives on the Field Sciences*, edited by Jeremy Vetter, 108–34. New Brunswick, NJ: Rutgers University Press, 2011.

Warming, Eugenius. *Plantesamfund: Grundtræk af den økologiske plantegeografi* [Oecology of plants: An introduction to the study of plant communities]. Translated by Percy Groom and Isaac Bayley Balfour. 1909. Reprint, New York: Arno Press, 1977.

"Washington: Mount Olympus Park." *Time*, July 11, 1938.

Watts, Lyle F. *Annual Report of the Chief of the United States Forest Service*. Washington, DC: Government Printing Office, 1947.

———. "Comprehensive Forest Policy Indispensable." *Journal of Forestry* 41 (1943): 783–88.

———. "A Forest Program to Help Sustain Private Enterprise." *Journal of Forestry* 42 (1944): 81–84.

———. "The Need for the Conservation of Our Forests." *Journal of Forestry* 42 (1944): 108–14.

———. *Report of the Chief of the Forest Service*. Washington, DC: Government Printing Office, 1945.

Wells, Gail. *The Tillamook: A Created Forest Comes of Age*. Corvallis: Oregon State University Press, 1999.

Wessels, Tom. *Forest Forensics: A Field Guide to Reading the Landscape*. New York: W. W. Norton, 2010.

———. *Reading the Forested Landscape: A Natural History of New England*. Woodstock, VT: Countryman Press, 1999.

"The Weyerhaeuser Idea as to Reforestation." *American Forests* 16 (March 1910): 194.

White, Richard. *Land Use, Environment, and Social Change: The Shaping of Island County, Washington*. Seattle: University of Washington Press, 1980.

———. *Railroaded: The Transcontinentals and the Making of Modern America*. New York: W. W. Norton, 2011.

Williams, Gerald W. *The Forest Service: Fighting for Public Lands*. Westport, CT: Greenwood Publishing Group, 2007.

Williams, Michael. *Americans and Their Forests: A Historical Geography*. New York: Cambridge University Press, 1989.

———. *Deforesting the Earth: From Prehistory to Global Crisis*. Chicago: University of Chicago Press, 2002.

Wilmking, Martin, and Jens Ibendorf. "An Early Tree-Line Experiment by a Wilderness Advocate: Bob Marshall's Legacy in the Brooks Range, Alaska." *Arctic* 57, no. 1 (2004): 106–13.

Winters, Robert K., ed. *Fifty Years of Forestry in the U.S.A.* Washington, DC: Society of American Foresters, 1950.

Woods, John B. "The Post-Code Status of Conservation." *Journal of Forestry* 33, no. 8 (1935): 710–12.

———. "Report of the Forest Resource Appraisal." *American Forests* 52 (1946): 413–28.

———. "Status of the Article X Joint Conservation Program." *Journal of Forestry* 33, no. 12 (1935): 958–63.

Worster, Donald. *The Dust Bowl: The Southern Plains in the 1930s*. New York: Oxford University Press, 1979.

———. *Nature's Economy: A History of Ecological Ideas*. Cambridge: Cambridge University Press, 1985.

Young, Jeremy. "Roots of Research: Raphael Zon and the Origins of Forest Experiment Stations." In *Fort Valley Experimental Forest: A Century of Research 1908–2008*, edited by Susan D. Olberding and Margaret M. Moore, 363–70. USDA Proceedings of the Rocky Mountain Research Station P-55, 2008.

Zon, Raphael. "Darwinism in Forestry." *American Naturalist* 47, no. 561 (September 1913): 540–46.

————. "Forests and Water in the Light of Scientific Investigation." In *Final Report of the National Waterways Commission*, appendix 5. S. Doc. No. 469, 62nd Cong., 2nd sess. (1912).

————. "Forests as a Factor in Flood Control within the Upper Mississippi River Valley." In *Relation of Forestry to the Control of Floods in the Mississippi Valley*, 85–113. H.R. Doc. No. 573, 70th Cong., 2nd sess. (1929).

[Zon, Raphael?]. "What Is Industrial Forestry?" *Journal of Forestry* 26, no. 1 (1928): 1–4.

Zon, Raphael, and Henry S. Graves. *Light in Relation to Tree Growth*, Bulletin of the US Department of Agriculture, no. 93 (1911).

Zon, Raphael, and William N. Sparhawk. *America and the World's Woodpile*. Circular of the US Department of Agriculture, no. 21 (1928).

————. *Forest Resources of the World*. New York: McGraw-Hill, 1923.

Index